ROADS TO POWER

ROADS
—TO—
POWER

Britain Invents the Infrastructure State

JO GULDI

HARVARD UNIVERSITY PRESS
Cambridge, Massachusetts, and London, England
2012

Copyright © 2012 by the President and Fellows of Harvard College

ALL RIGHTS RESERVED

Printed in the United States of America

Library of Congress Cataloging-in-Publication Data

Guldi, Jo (Joanna), 1978–
Roads to power : Britain invents the infrastructure state / Jo Guldi.
p. cm.
Includes bibliographical references and index.
ISBN 978-0-674-05759-3 (alk. paper)
1. Infrastructure (Economics)—Great Britain. 2. Roads—Government policy—
Great Britain. 3. Transportation and state—Great Britain. I. Title.

HC260.C3G85 2011
388.10941—dc22   2011014354

To John Stilgoe and James Vernon

CONTENTS

    Introduction: The Road to Rule   1

1. Military Craft and Parliamentary Expertise: The Institutional Evolution of Road Making   25

2. Colonizing at Home: The Political Lobby for Centralizing Highways   79

3. Paying to Walk: The National Movement against Centralized Roads   128

4. Wayfaring Strangers: Mobile Communities and the Death of Contact   153

    Conclusion: The Necessity for Infrastructure   198

Notes   215
Acknowledgments   289
Index   291

ROADS TO POWER

INTRODUCTION

The Road to Rule

> Could the England of 1685 be, by some magical process, set before our eyes, we should not know one landscape in a hundred or one building in ten thousand. The country gentleman would not recognise his own fields. The inhabitant of the town would not recognise his own street.
>
> —THOMAS BABINGTON MACAULAY, *The History of England from the Accession of James II*, 1843

In 1726, the roads of Britain were mire and muck. A few cobblestoned streets in well-off villages punctuated long stretches of dirt track between towns. Rain-soaked wheel ruts and eroding banks made long-distance travel impossible for considerable periods of the year. Occasionally a peasant dug a hole in the middle of the road to obtain mud to make bricks. If the hole was disguised by rainwater, a traveler's horse could disappear into it. The courts had only recently declared this practice remediable.

By 1848, the road system consisted of forty-foot-wide highways of level gravel that extended to every village and island in the nation. The tiniest pebble of the road had been measured by the hands of out-of-work women and children, watched by appointed surveyors who sent paper forms testifying about their repairs back to engineers in the nation's capital. Networks of strangers traded goods and intelligence by post, stagecoach, and rail, and working-class radicals, traveling on the same roads, plotted the possibility of equal voting rights for all. Houses had been torn down to make room for the roads, and officials in London combed through piles of reports about the regulation of distant surveyors. Much had changed, and contemporaries wondered about the nature of the transformation. Was it a new era for sociability? Or was it a new era of surveillance, where no man's property was safe from the expanding state? Had roads finally fashioned a nation of the four kingdoms of England, Scotland, Ireland, and Wales? Or had the politics behind the roads, as some thought, pitted Celt against Englishman, race against race, in a constant competition for funding?

These are not unfamiliar questions in the era of interstate highways and the Internet. In the first years of the twenty-first century, China connected its major cities by a thousand-mile rail network of hundred-mile-an-hour trains, relying on infrastructure to engender economic expansion. In the United States, legislators argued about "bridges to nowhere" paved at taxpayer expense. Parents' groups warned about the dangers of strangers on the Internet. A "digital divide" separated children in poorer schools from the informational economy of rich ones. Optimists predicted that cheapening technology would eliminate differences of income and access, and drawing upon a hypothesis of unlimited technological innovation, futurist Ray Kurzweil prophesied the end of conflict as technologies connected not only laptops but also brains. In the modern world, we lust for connectivity and fear its cost. We debate the levels of surveillance necessary, and we fear the ways in which it may go awry.

In 1852, essayist Michael Angelo Garvey conjectured in his book *The Silent Revolution* that because of road transport, divisions would disappear and all the peoples of the world would soon speak one language. Every aspect of Garvey's world had been transformed by the coming of road and rail. First the stagecoach and later the railroad carried with them standardized time, and regional clock towers were aligned not with the sun but with Greenwich. Newspapers began to reach the farthest islands of Scotland, and post offices with regular delivery times appeared in the smallest towns in Wales. Businesses spawned franchises along the corridors that supported this new traffic, and by the 1830s, standardized inns and packhorse syndicates could be found in every major city in Britain. The broad distribution of such services and resources regularized so many aspects of daily life that one might reasonably expect with Garvey that strangers would soon come to share a single set of ideas.

But to what extent were strangers brought together by advances in transport? Middle-class persons in a stagecoach, moving from inn to inn, equipped with pocket watches and shepherded by maps and guidebooks, had fewer practical reasons to interact with strangers. Interactions between rich and poor were reoriented around privileges of access to transport developments. Those who could afford to make use of new transportation did; those who could not afford it continued to get around as they had for generations—on foot. Privileges once enjoyed by the upper classes became the norm for middle-class travel. Isolation had the effect of stigmatizing foot travelers, now visibly marked as the wandering poor. Prejudices arose against mobile communities of journeymen. Methodists were marginalized, and their messages were ignored. "Small talk" became a middle- and upper-class strategy for discerning just which travelers might undesirably interrupt the ease of a solitary, independent journey. Although these travelers moved along the same routes, they occupied increasingly isolated spheres. Rather than being homogenized by infrastructure, their routes and meeting habits were developing new forms of segregation.

Nor did roads cause the end of political struggles. Rather, the opposite was true. Building the roads had involved Parliament in unprecedented spending, and parties began to realign over the question of where the roads should go. Entire regions, such as Scotland, might have their hopes for prosperity dashed by funding disputes in London. The continued prosperity of rich and poor regions depended on decisions made far away in Parliament. By 1830, Scotland, Ireland, and isolated farmers in the north of England successfully lobbied for government roads to connect the entire nation, and some nine hundred miles of road were built. Meanwhile, the English gentry inaugurated a libertarian revolution that would cut off government-funded transit altogether, and within twenty years, the northern roads, formerly "the best in Europe," had been reduced to crumbling tracks. Rich and poor regions were pitted against one another as new political fault lines opened over issues of access to infrastructure.

In the modern era, fighting over the design of roads has become a valuable way of achieving larger political goals. Conflicts over public housing, eminent domain, levees, and infrastructure funding divide modern America as directly as conflict over the roads divided nineteenth-century Britain. As modern people built roads, then railroads, and then modern cities, bus lines, subways, and public housing, competition over access to resources followed. The exile of poor persons to the edge of the city and the exile of poor regions from the system of infrastructure—these are things we fight about in the modern era.

Insofar as political activity is engendered by changes in infrastructure, I will consider it the politics of an infrastructure state, for to be political

INTRODUCTION

in this era is to raise questions about how that state works. In the infrastructure state, governments regularly design the flow of bodies, information, and goods. Modern governments in developed nations have mediated the relationship between individuals and infrastructure technology for so long that the role of the state in designing ports, sidewalks, and bus lines is nowadays taken for granted. It was through state activity, in the form of copyright and the broadcast spectrum, that information assumed the shape of property. By regulating railroads, states taught technocrats to cooperate; with libel and tort, states made the users of technology liable for negligence. In modern cities, government engineers design large-scale transport infrastructure, as well as street lighting, sewers, schools, and libraries. Governments directly orchestrate the social structure of neighborhoods through housing codes, health codes, zoning laws, preservation schemes, and mortgage restrictions. There are few aspects of the built environment that have not been shaped by government.

I place the origins of the infrastructure state during the period between 1726, with the military survey of Scotland, and 1848, when the British government abandoned its roads. Since at least the early nineteenth century, historians have tried to understand this period and its relationship to infrastructure, frequently calling attention to the role of bridge designers and civil engineers like Thomas Telford, whose efforts were glorified as early as 1861 in Samuel Smiles's textbook study, *Lives of the Engineers.* But such studies tell us little about the early initiatives to build modern roads, and even less about how infrastructure came to be a subject of contestation.

In Britain, eighteenth-century surveyors for the military, the mail, and development brought remote islands under the surveillance of government. Parliamentary road building required common adjudication of land, as well as collective management over sometimes thousands of miles. The centralization of control took place under the guidance of the state. Government spending, state-appointed experts, and an army of clerks and surveyors oversaw the coordination of standardized designs for roads and the circulation of standardized forms specifying the details of construction, down to the exact dimensions of each piece of gravel on every road surface in the farthest village of each tiny island within Great Britain. Across the nation, battalions of unemployed poor persons raked that gravel, while elderly women sat by the roadside, sorting the gravel according to size. Tollbooths, way markers, sidewalks, and foundations were all suddenly standardized, a process that required the reorientation of labor and capital to

INTRODUCTION

new centralized management. Hundreds of thousands of pounds were expended on monumental bridges like that across the Menai Straits, the first suspension bridge in the world, henceforward visited by coachloads of tourists who came to sketch the vast work of steel, surrounded by forest. These tourists bore witness to the simultaneous birth of modern infrastructure and modern bureaucracy.

Transport and the state had not always belonged together. In the ancient world, the paved roads of government were an anomaly in a landscape traversed by footpaths. The earliest long-distance trade routes extended across the borders of one of the earliest economic divides, that of the Eurasian steppes, where pastoral nomads traded with farmers in the valleys below. By the first millennium B.C., these sporadic trade networks flowed into the silk roads that stretched from China to the Mediterranean. In the following centuries, the paths of silk traders shifted with the tides of trade and the threat of war. They connected with caravans of camels coming out of Africa and sea routes charted by the skilled navigators of the Indian Ocean. Over these routes were carried stories, religions, diseases, and political systems.[1] It was the route, not the road, that carried commerce and religion.

Roads were something else: a mechanism for government and a tool for the travels of soldiers, but very rarely an artery for trade. The ancient roads of Persia and Rome connected only destinations useful to administrators; they carried soldiers and judges but rarely oxcarts or packhorses. The military roads built under Henri IV and Louis XIV carried soldiers across isolated peaks, but they too had little impact on the course of trade.[2] The ancient highway was the realm of the state. Trade happened elsewhere, where government was limited and the pathways were plowed by the exigencies of commerce.

The travelers of this ancient world met on the tracks of animals and the ports of the sea. Trails wide enough for packhorses but not imperial vehicles carried tin over the peaks of ancient Britain and France, repeating overland trading routes that converged on pre-Roman Marseille, where tin was exported to Phrygia in return for wine and salt. Provincial traders in Roman Egypt traveled in a world dominated by the customary routes of the Nile. Medieval friars and journeymen found their way from town to town by asking fellow travelers for directions. With the expansion of Chinese and Indian trade and innovations in maritime technology between 1500 and 1750, the sea became the primary medium for a global chain of routes, linking

INTRODUCTION

the world's economies together and superseding many forms of overland trade.³ None of these exchanges depended on regularly maintained, centralized infrastructure of the kind witnessed in ancient Rome.

That division—government road and merchant route—collapsed beneath the infrastructure revolution. In the seventeenth century, city-states began organizing their collective wealth around the provision of canals, the first government-built corridors for carrying commodities rather than soldiers. By the nineteenth century, infrastructure had taken the form of state-designed sewers and slum-clearance projects, tools of social as well as civil engineering. They had created a new world, as geographer Tim Ingold has observed, where government involvement in the built environment was increasingly experienced as a given.⁴ The actors who reshaped exchange and sociability were no longer mere mule drivers and merchants. In the modern world, travel was governed by experts and statesmen. Through government highways, canals, and railroads, and later, through bodies that ratified the courses of industry, the bureaucrat became the judge of where trade should flow, who undertook the cost of carriage, and who could participate in the decisions involved.

Thomas Rowlandson, "Post Office," in A. W. G. Pugin, *The Microcosm of London* (London: T. Bensley, printer, 1808).

INTRODUCTION

In France, Britain, and Prussia, early modern governments harnessed engineering on behalf of political security. Advanced military fortifications secured national borders in the era of Vauban. The Canal du Midi was constructed to link politically fissile Languedoc with the north of France between 1666 and 1681. Britain's military roads of the 1740s inspired Napoleonic corridors that channeled soldiers into Italy and France between 1798 and 1815.[5] Early modern states built infrastructure, as had ancient Rome, around the logic of conquest.

Elsewhere, infrastructure was bent not to imperialism but to the interests of small communities looking to increase their agricultural productivity and diminish the cost of transport. In the seventeenth century, Dutch engineers began poldering swamps into arable land and administering networks of small canals. Their techniques were exported to East Anglia, where Dutch windmills transformed the fens from swamps into farmlands. River navigation brought trade to inland towns, and canals linked elevated mines with external ports. Turnpike trusts appeared in Britain by the 1640s, through which local towns reorganized their transport needs after Henry VIII's 1536 dissolution of the monasteries.[6]

Daniel Defoe described the landscape of the late seventeenth century as a patchwork of economic and political experiments. In the stretch of the Great North Road that connected Hatfield to Stevenage and Bugden, travelers encountered a passage where enterprising locals had "placed Gates, and laid their Lands open, setting Men at the Gates to take a voluntary Toll, which Travellers always chose to pay, rather than plunge into Sloughs and Holes, which no Horses could wade through." This network of tiny, start-up turnpikes through every field allowed travelers to skirt "a most frightful Way."[7] Defoe recorded a dozen such examples of start-up turnpikes, from small, gated trails to great, paved causeways. Trade along turnpikes began to unite ancient market towns into regional trading blocs.[8] These small-scale transformations enriched the few and created archipelagoes of industrial luxury within a sea of premodern agriculture. In the few villages wealthy enough to pave their roads or light their streets, the traveler glimpsed what modernity might resemble.

Such an apparition of abundance was enough to inspire new theories of wealth. Those who observed Dutch dikes and English turnpikes drew inspiration from the fact that small communities profited from collaborating on infrastructure. Political philosophers such as Bernard Mandeville, the Marquis de Condorcet, Anne-Robert-Jacques Turgot, and Adam Smith began to examine collaboration among self-interested individuals, speculating about how larger governments could transform trade on a national scale by legislating conditions in which natural economic laws would generate prosperity for all.[9]

Systems thinkers reasoned that governments that collected information could act so as to speed the economy as a whole. Chief among these envisioned reforms was the building of infrastructure. Turgot, the French finance minister, recommended that a sound government rationalize its tax codes, remove tariffs on internal trade, stabilize the currency, and provide basic infrastructure. Such measures would extend the possibility of economic participation to all the inhabitants of the nation, allowing country farmers and city guild members to trade on equal terms, freed from medieval restrictions and fines. Sound roads, canals, and ports would solidify these freedoms, bringing possible trading partners together at a fraction of the cost of medieval travel.[10]

Such thinking proved inspirational to early liberal governments. In Britain, the canal boom of the 1760s was protected against monopoly by government intervention. In America, state and local governments protected and subsidized the corridors of canals, rails, and roads that opened the frontier between 1815 and 1861, with the federal government committing some $300 million to internal improvements. By the mid-nineteenth century, most modern states were experimenting with a variety of public works—river navigation, canals, railroads, lighthouses, ports, and shipping canals—but all modern states were building roads. As in the writings of Garvey, roads promised to unite and stabilize. Unlike canals, ports, and river navigation, road networks could reach all potential market participants, regardless of geographic contingency. By cutting across watersheds and into poor peripheries untouched by private rail, roads appealed to modern states as a leveler that put rich and poor regions on an equal playing ground. Roads even enhanced access to canals and rail and, on the scale of small streets, remained attractive even where there were trains, with the result that horse-drawn transport continued to expand dramatically during the era of the railway until its replacement by the motorcar.[11] Whether for reasons of imperial conquest or economic advancement and competition, modern governments were all but forced to invest heavily in infrastructure.

Between 1726, the year the British government began investing in its roads, and 1848, when state control had reconstructed the entire road system, the British road system changed from a scattered set of tracks and

INTRODUCTION

pavements into a carefully regularized network. Contemporary atlases suggest the scope of the change. A map from the year 1726 marked particular routes through major cities but lacked detail and calculation. In that year, only an occasional road atlas bothered with a diagram of how Britain's many routes connected into one whole. The popular image of the road network was a suggestive diagram that charted nine to fifteen major axes of travel through Britain without specific information about the numerous variations of each route. Road atlases concentrated on the specific landmarks necessary to navigate particular routes: hills, mansions, villages, and turnpikes. The routes as they were diagrammed—that is, as straight lines on the page—corresponded little with real turns in the landscape. Thus the road network in 1726 existed only in the imagination of the cartographer; the traveler was expected to ask other travelers, innkeepers, or passersby for directions, and the path suggested would vary with the wetness of the season and the condition of different routes. The atlas itself helped little: too large and cumbersome for travel, or hand painted in colored inks on expensive vellum, it was more of a luxury item than a tool.[12] Additionally, it was an English rather than a British atlas. Scottish roads, if they existed, were assumed to be outside the interest of the English traveler.

A road map of 1848 was, by contrast, an accurate documentation of a new road network. The map's neatly engraved, precisely delineated traces of the roads' shapes, diversions, and connections were the product of the Post Office commissions that after 1785 had paid private surveyors to accurately tabulate the distances between each hamlet and the dome of St. Paul's.[13] A precise netting of minute lines filled the map where only broad arteries had appeared in 1720. The geographic borders of the map expanded to include Scotland, which was filled with its own neat lines connecting the entire nation through the farthest Highlands. By the 1790s, state subsidies for road building and surveying had launched an explosion of mapmaking, and detailed forty-centimeter folding maps were available on paper or linen for as little as two shillings.[14]

Immense variance in the quality of early eighteenth-century British roads reflected a decentralized system where each parish was entirely responsible for the shape, direction, and quality of its own roads, and little technology or incentive existed to improve them. The traveler in 1726, the year the British government began investing in its roads, would have journeyed through a series of disconnected islands of road governance. At that time, even London offered only "dirty, or hard-paved rattling Streets." The

John Ogilby, *The Traveller's Guide; or, A Most Exact Description of the Roads of England* (London, 1712).

John Arrowsmith, *The London Atlas of Universal Geography* (London: Published by J. Arrowsmith, 1842).

wealthiest towns, such as Kingston, a market town within sight of Hampton Court, had only "stiff Clays" formed into "very bad Roads."[15] Defoe found perfectly cobbled streets in the rich lace-making town of Honiton in Derbyshire, but "deep, dirty, and unfrequented" lanes serving the poor spinners of Spitalfields within sight of London.[16] Fewer than 2 percent of all parishes were spending any money on their roads in any given year.[17] Faced with these variations, the traveler's expectations of speed and route planning had to be negotiated as he met with conditions on the ground: Defoe's journey virtually ground to a stop on the rutted roads near Kingston, and he nearly lost the track entirely in the uncharted "houling Wilderness" around Yorkshire.[18]

By 1843, the year in which Thomas Macaulay canonized the transport revolution in his *History of England,* the traveler who crossed these same distances would have experienced a different transport infrastructure altogether. Roads stretched into the mountainous and swampy peripheries of the island, clinging to the mountainsides of the Highlands, carried by long bridges through Scottish canyons, buoyed by thick foundations over Yorkshire's swampy moors, and embanked over the shifting sands of Wales's tempestuous shores.[19] Wide highways crossed long stretches of mountain waste where no human settlement existed to maintain them. The most impressive bridge in the country, the gigantic suspension bridge across Menai Straits, connected the backwater harbor town of Holyhead to the isolated and mountainous shore of western Wales. England and Wales together had 119,527 miles of roads that yearly consumed £1,600,000 in parish rates and £1,097,000 in turnpike tolls, of which up to £570,490 was directly managed by Parliament.[20] Those yearly incomes had been supplemented by a dozen additional grants from the Treasury of sums up to £759,718 for several hundred magnificent bridges and 2,600 miles of new roads connecting England to underdeveloped regions of Scotland and Wales.[21]

The new highways were quickly flooded with traffic. Even remote highways could be busy; one was described by a German visitor in 1782 as "more alive than the most frequented streets in Berlin."[22] Equally impressive was the physical infrastructure beneath travelers' feet. The whole of this road network consisted of a single continuous surface, a bed of gravel, flint, limestone, whinstone, or pebbles ten inches thick and twenty feet wide. The stones, broken into even shapes and washed clean of loam and sediment, were piled into a sloping mound, raised in the middle, so that

INTRODUCTION

water would drain into ditches on either side of the road.[23] Hedges and trees were manicured so they neither obstructed nor shaded any part of the road for a full thirty feet on either side of the highway. Wide footpaths raised above the curb accommodated pedestrians on both sides for the road's entire course, and the whole of its length was punctuated with milestones and guideposts.[24] Local variations in road surfaces vanished with the Metropolitan Improvements Bill of 1817, the rulings of King's Bench in the 1820s, and the Highway Act of 1835, which forced the parish surveyors of highways to comply with parliamentary statutes about the shape and management of roads.

The roads thus constructed marked a significant departure from investment in transport infrastructure to that point. In 1726, roads were secondary to other forms of transport, such as ocean and river shipping, and most maps stressed the water-bound arteries of communication in preference to the inland roadways that connected village to village through the countryside. Defoe's journey described major military building around Britain's ports and harbors, but the state had no presence at all in the care of the roads. Drawing into Tilbury past the marshes of Essex, he described the immense fortifications, the Tilbury Fort, planned as "the Key of the City of London," its esplanades and bastions embanked on iron-girded posts lodged in the mud. The still-unfinished fort had been armed with 106 cannon to protect the Thames estuary from invasion. There were also the four lighthouses beyond Yarmouth, "flaming every Night," the "battery of guns" along the shore, and the strong harbor at Harwich, "so firm that neither Storms nor Tides affect it." Each of these immense structures was the product of "many Years Labour, frequent Repairs, and a prodigious Expence."[25] The Harwich ports had sent one hundred men-of-war forth to clash with the Dutch in explosions of guns, fire, and smoke, and the defenses had still been under construction as recently as the 1680s. The coastal landscape along which Defoe passed was punctuated everywhere by icons of the military-fiscal state.

Seventeenth-century river and coastal improvements had provided England with a network of water transit focused on the manufacturing districts of the north and the iron-producing territory of Wales.[26] So dominant was water transport in 1726 that many works of geography did not even bother to include the roads in their description of the island. Thomas Cox's British atlas of 1720–1731 included maps of Britain's seacoasts and waterways but no map of land routes beyond the county level.[27] More typically, the routes were simply schematized down to the major corridors of access. A 1720 edition of William Camden's *Britannia* included a recent map by Robert Morden that showed only two routes through the whole of Northamptonshire.[28] The seventeenth-century focus on river ways and harbors extended into eighteenth-century canal and port development. Canal and port industrial corridors were characterized by their impact on narrow strips of geography at the ocean's shores. Such developments left inland villages radically disjointed and altogether unfamiliar with one another.

In the course of the next hundred years, that pattern of investment changed drastically. The state began a series of new building projects in 1726 that would directly alter the experience of future travelers. Road building in Scotland pioneered modern practices of surveying, labor management, and road engineering, which later fed the development of roads by turnpike trusts. Military funding built the first hundred-mile stretches of roads where Thomas Telford constructed his ambitious bridges. Government concerns with security of frontiers and the mail drove the construction of Britain's interkingdom highways, connecting London with Edinburgh and Dublin (via Holyhead), and John Loudon Macadam and Telford were deployed to erect bridges and embank mountain roads in Scotland and Wales. These state activities transformed the shape of the nation, connecting its remote corners, welding local variations in road surface and management into a single systemic unit, and seamlessly working to create a uniformly paved structure across the entire island. By 1848, the military investment so visible along the coast had been largely redirected to the nation's interior.

These changing forms of investment called for commensurate changing forms of bureaucracy. The number of surveyors and engineers responsible for roads expanded. Military surveyors trained under the military Ordnance Survey and civilian surveyors elected by the parishes were brought under a new regime of centrally organized forms and plans. By 1835, Britain had organized a system of surveyors overseeing not only major highways but also smaller roads and streets. In France after 1840, the Corps des Ponts et Chaussées went from managing strategic highways to supervising the nation's minor or "vicinal" roads, and new roads were designed and managed by engineers trained at the École des Ponts et Chaussées. The peripheral road network was paved in Britain between 1815 and 1835; in France, between 1840 and 1870; and in America, between

1870 and 1930. By 1870, Britain, France, Prussia, and America all had modern systems of pavement and management where manuals, forms, and bureaucratic hierarchy ensured that the smallest pebble on the farthest island was scrutinized by the state.[29]

Roads also allowed a concomitant expansion of the post, together with the number of clerks, drivers, and postmasters responsible for its operation. Britain pioneered the post-coach system of passenger and mail transport with the version set up by John Palmer after 1784 and perfected into the Penny Post in 1840. In France, post coaches set up under Louis IX were turned into a full rural delivery system between 1830 and 1870. And in America, the building of federal post offices in most cities and towns during the 1860s and 1870s was followed by the institution of rural free delivery in 1891. In Britain, France, and America, official post offices appeared in small towns and major cities, their temple fronts reminding locals of the everyday presence of government.[30] Expanding infrastructure constituted an unprecedented expansion of bureaucracy, foreshadowing later government undertakings in the name of welfare and public health.

Infrastructure united the nation in terms of mail, commerce, and travel, and it promised to do more. Promoting the highway system in Parliament, the Scottish landlords urged the power of roads to transform a nation, to unite separate ethnicities (like Celts and Englishmen) into a single people, to promote intermarriage, to overcome linguistic divides, and to quell the risk of military rebellion. The roads seemed to promise even greater forms of unity for the nation, a valuable tool for governments in the service of peace.

The origins of modern infrastructure have been frequently misunderstood in ways that mask the role of the state. Since at least 1913, when Sidney and Beatrice Webb published their *Story of the King's Highway*, modern historians have tended to draw attention to the expansion of roads during England's eighteenth-century turnpike boom. Private road-building projects in the 1740s were followed by private canal building like the Duke of Bridgewater's canal to his mines outside Manchester, opened in 1761. James Watt's steam engine was developed between 1763 and 1775, and by 1832, England became home to the world's first railroad, funded by private investors in Birmingham. Spurred by this success, British investment and British railroads spanned the globe, backing private companies in the United States and South America.[31]

INTRODUCTION

This story emphasizes the role of private individuals: entrepreneurs and investors like the Duke of Bridgewater and inventors like Watt. From the publication of Samuel Smiles's classic *Lives of the Engineers* (1861) onward, these names have filled standard historical textbooks as the makers of the modern world. As an economic historian like David Landes or Joel Mokyr might explain, early investments and political stability produced a climate where new inventions could proliferate. In England, where the state was relatively absent (as the story goes), private turnpikes and canals could generate handsome profits, and inventors began to flourish. Stories about entrepreneurs and inventors suggest many lessons, but especially the role of private enterprise and individual imagination in creating the modern world. Government has very little place in this story; rather, the absence of government is what allowed the Industrial Revolution to happen.[32]

The story changes greatly if we look back a mere forty years before the opening of the Bridgewater Canal. Then, military road-making technology was introduced to England from the Continent and was refined further as military engineers perfected their arts in the winding mountains of the Scottish Highlands. The technology behind the transport revolution, including the siting, foundation making, and labor management of turnpikes, originated in military road building in the early eighteenth century. Britain's military roads, constructed between 1726 and 1750, served the troops during the period of military control of Scotland as English soldiers began policing the Highlands and confiscated the land belonging to the rebel lairds who had recently marched on Newcastle. Roads ensured that the English army could police the rebel state. More than nine hundred miles of road were paved, connected by one thousand bridges, which were maintained at public expense through the 1790s at a cost of some £5,000 yearly.[33] It was hoped that they would also increase industry in the areas through which they passed, but this was a secondary concern; they led straight to the glens where rebel lords dominated, passing few centers of population or industry. They were funded from the same taxes that fueled other military expenditure, and they formed a major laboratory for the invention of new siting, surveying, and paving technologies in difficult territory. At the end of the hostilities in Scotland, English veterans returned to their homes, armed with knowledge of how to make roads. New technologies of road building soon began to spread over the ancient system of parish roads as the veterans sold their services to local governments.

INTRODUCTION

Turnpikes spread over the majority of the English interior between 1740 and 1760 and inspired new conversations in England about the necessity of infrastructure. Turnpike trusts, which shifted the burden of road improvement from local householders to the broader class of travelers as a whole, penalized long-distance travelers and erected an expensive series of barriers that inhibited national trade. From the perspective of the parish government in Newcastle, turnpikes brought military technology to the towns at an affordable cost. But from the perspective of the Edinburgh trader who wished to carry goods overland to trade in London, the hundred turnpike trusts between him and the market made trading prohibitively expensive.

These concerns about trade provoked an unprecedented reorganization of the government. Through the crusading of Scottish landlords, Parliament was pressed into the task of building roads for the entire nation. Within ten years, the turnpikes were superseded by an interkingdom highway system. These roads were overseen by centralized engineers, coordinated by Parliament, and designed to connect London with its colonial capitals. They brought roads to areas of the nation where previously there had been none. New legislation wrapped the whole of British streets and byroads under parliamentary control.

The sheer scale of the undertaking was astounding. Four major projects—the Highland Roads and Bridges, the Holyhead Road to Dublin via Holyhead, the Great North Road between London and Edinburgh, and the knotting together of these routes where they crossed in London's streets—aimed to wield the power of the nation-state to reshape Britain's transport system into a single organ for assimilating the manpower and wealth of its Scottish and Irish colonies.[34] Between 1803 and 1835, the British Treasury financed the building of roughly seventeen hundred miles of smooth roads that radiated from London to the north and the west across expensive embankments and suspension bridges. Government building extended the commercial network of post coaches and inns to the farthest corners of the nation. The Highland Roads project developed hundreds of miles across the farthest, underdeveloped Scottish periphery, and the Holyhead Road and the Great North Road together connected London to Edinburgh and Dublin, making possible the more rapid and powerful exchange of information and trade between England and its former colonies and effectively transforming the map of the British market. As infrastructure bolstered and evolved trade, disputes arose over entitlement to infrastructure access.

Old kingdoms were reconstituted into new political identities over disputes about who should fund infrastructure. As centralizers and localists, or colony and metropolis, Britons battled one another for control over and access to the roads. The expanding horizons of trade opened up deeper struggles than ever before, struggles over the shape of trade, over who would own it, and over whether local or centralized government would better manage it. Peripheries like Ireland and the Highlands demanded that centralized government pay for their roads, which they could never afford by themselves. English counties protested with a defense of local government that would save them from the cost of underwriting highways to Scotland. With each successive generation of road building, new models of funding were pioneered and criticized. Each theory of the market was identified with the particular regions that stood to benefit or lose by its implementation.

This competition among regional lobbies drove up the scale and cost of development spending. By the 1820s, the price of connecting the periphery provoked a libertarian backlash against parliamentary spending. By 1835, libertarian victory was in place; no further plans for national roads were approved until the twentieth century. Parishes continued to control the vast majority of Britain's roads until the 1860s, when their authority was gradually assimilated into larger county or regional highway boards. As a result of this struggle, finances for most of Britain's major arteries remained local until after 1909, when the Road Fund provided an alternative, centralized subsidy. Nineteenth-century plans for centralized control of the roads were realized only after the Trunk Roads Act of 1936.[35]

Infrastructure politics proved divisive in Britain and beyond. The possibility of centralized revenue, crystallized around opportunity and expense in the highways, creates a climate where competing groups clash with one another to control the shape of future development. In the infrastructure states of nineteenth-century Prussia, France, and America, regional lobbies similar to Britain's battled for funding. In Prussia, draining and port-building projects under Frederick the Great forced poor peasants from their land. In America, farmers along the American and Mississippi rivers battled one another for the largesse of the Army Corps of Engineers.[36] Conflict over the government's role in infrastructure is one of the primary features of the infrastructure state.

The arguments once made by regional interests have been defended so often, so persuasively argued for, and so abstractly theorized that we now

INTRODUCTION

typify them as the reflection of political ideology. Socialist France pays for its trains, while libertarian Texas prefers toll roads. Journalists and even historians write about road building as a conflict between the state and the free market. Libertarian economists like Edward Glaeser argue for the devolution of roads to private control, regardless of how remote or poor a region is. In reality, the divisions at stake are frequently deeper than political ideology suggests. Poor regions and rich fight not over abstract political ideals, but over the right to participate in the market.[37]

In the era of the infrastructure state, conflict is inevitable because building, although expensive, is necessary. Without state building, economies never expand to a national scale, peripheries are left behind, and the poor cannot afford to participate in the market. Rich regions and classes inevitably protest the expense of a national system. Infrastructure pits region against region, experts against the people, and class against class. It produces and informs the identities and divisions that characterize politics in the modern era.

It can hardly be surprising that such heated conflicts over infrastructure would extend to broader cultural issues. As the roads spread, the state became involved in the regulation of public space as well as public works. New laws required wide sidewalks for pedestrians. Parliamentary legislation on traffic, garbage, and snow generalized the authority of local communities to control their public spaces. Individuals were required to pay vehicle taxes to the Stamp Office and to carry license plates on every wagon. These regulations were also bound up with attempts to police the social order. As the civil engineer James Loudon Macadam explained it, properly administered roads could eliminate poverty. Not only would highways bring prosperity to distant villages, but able-bodied members of the out-of-work poor could also be employed in breaking rocks and sorting gravel. This passive system of eliminating poverty by road building could be supplemented by new systems of surveillance and control. Vagrants had been whipped in public since the mobility crises of the sixteenth century, but by 1810 new treatises contemplated a systematic, national system of policing their movements. Every homeless person in the entire nation who asked for relief would be assigned a profile. These persons would then be carted to the areas where jobs were available. If any authority in the nation discovered them asking for bread after they had been offered work, they would be thrown into prison. For local officials, the expansion of infrastructure suggested how the state could collect and sort information to police the everyday movements of the poor.

Theories about policing poor travelers were spoken of in the same circles where intellectuals imagined that transport would create new levels of commerce and sociability. Eighteenth-century observers noted how city centers promoted the indiscriminate mingling and sharing of ideas, and Kant theorized that travel itself instilled the value of "cosmopolitanism" in individuals and nations.[38] Before the era of political conflicts over infrastructure, the social experience of travel encouraged openness among people. Defoe advised travelers to do as he did: find their way with the help of strangers.[39] Strangers were easy enough to come by. Defoe came across a "great concourse of Travellers" on the old Watling Street from London to Warwick.[40] He found them intermingling at village fairs and chatting at inns. When they traveled together, they traveled in intimacy on roads four to twelve feet broad.[41] Defoe stopped to speak to the innkeeper of each inn on his route. Such intimacy generated everything from terrifying encounters with highwaymen to casual conversations between fellow travelers exchanging directions.

Nursing such intimacies with strangers was even more important as one moved farther into the periphery. Defoe recommended that travelers avoid Scotland altogether unless they could use their connections to gain "the countenance of the Gentlemen, and the chiefs, . . . and to be recommended to them from their friends from one to another, as well for Guides as for safety." He complained about a variety of potential grievances facing the traveler. In Harwich, innkeepers accustomed to fleecing Dutch traders were "far from being fam'd for good Usage to Strangers," he cracked, "but on the contrary, are blamed for being extravagant in their reckonings, in the Publick Houses." Yet the gentlemen he met in Galloway were "particularly very courteous to Strangers merely as such, and we received many extraordinary Civilities on that only Account."[42] Custom and economic opportunity structured the kind of exchanges Defoe encountered, but by the nineteenth century these forces were being displaced by others.

Travelers in 1848 would have as little pretext for such discoveries as they had need for conversation with strangers. By 1848, the major roads were thirty feet wide.[43] These highways consolidated the divergent pathways of earlier routes into major corridors that lacked the intimacy of earlier roads.[44] Maps, brochures, printed travel guides, and parliamentary signposts now saved the traveler from asking strangers for directions.[45] The middle-class tourists and traders who followed their instructions traveled by post coach to inns where they were cushioned by people of the same class and background against the shocks of unfamiliar surroundings.[46] The few tourists who longed for encounters, like Anne MacVicar Grant, visiting the Highlands in the 1770s, despaired of other tourists, "insipid aliens," incapable of interacting with either strangers or landscape, who, on the rare occasions "when they walk," went no farther than "the hard gravel road, to get an appetite."[47] If anything, the traveler in 1848 had a less social experience, was judged and criticized at a distance by others, and was encircled by a fog of tourist guidebooks and naïveté.

Other meetings were being structurally formalized and controlled by the intervention of the state. Since the 1690s, the military-fiscal state had deployed droves of excisemen, soldiers, and post-office surveyors on the provincial roads of counties throughout the three kingdoms. As these groups of travelers began to recognize one another as members of a community, they recorded an understanding of themselves as a people apart, united by a common experience of travel and mutual responsibility. After the 1740s, soldiers on rotation, Methodists on circuit, and artisans on tramp grouped into mobile tribes, traveling together for protection and sociability on the road. Shared stories and songs, drawing on the biblical language of pilgrimage, united individual travelers into a larger community imagined as a people apart. By the 1790s, fear of revolution on the Continent prompted British elites to contemplate turning powers of infrastructure regulation against the poor travelers who carried radical ideas. Conflicts emerged over vagrancy, homelessness, and sanitation, and in reaction, new laws were passed restricting the rights of access to the street of some and refining the experience of others.[48] Increasingly, middle-class communities passed regulations that penalized poor travelers. Faced with suppression from above, the cosmopolitan tribes united by the roads fell apart. Methodist field preaching was abandoned, journeymen travelers were suppressed, and vagrants were taken to the workhouse. The era of political conversation on the public highway was over.

By 1800, middle-class Britons frequently argued that they inhabited a nation that had been unified and civilized by roads. The reality was quite the opposite. By 1848, poor pedestrians and the polite travelers of highways and guidebooks moved in different worlds. Rarely, if ever, did these classes in divided Britain interact on the nation's highways. Rather, it was in the course of the transport revolution that strangers became so deeply separated.

In 1726, the British state began to engineer the infrastructure necessary to unite England with its hinterlands and former colonies, but by 1848, infrastructure had launched the creation of alternative communities within the nation's borders. Radical journalists and agitators for local self-government envisioned a Britain of parish governments in which the roads were returned to community rule. Methodists and union leaders imagined that they belonged to another nation, Jerusalem, overlapping but at odds with British identity. Tensions over infrastructure forced regions to compete for funding and launched a libertarian revolution that dismantled much of the state bureaucracy.

The infrastructure state also inaugurated an era of public regulation where policing public space became the surest way of establishing local control. New laws systematically demonized and penalized mobile communities of working-class travelers, paralyzing the vision of a nation united by its roads. The poor traveler became, like the poor region, a category despised by the English county.

In debates over highways, rail, the post, and broadband, modern persons often believe that information and technology unite distinct peoples into one body. Britain's transport revolution was the first major public episode of uniting the everyday journeys of commerce into a system of high technology and top-down control. But Britain's experience was not one of connection. By 1848, what Britons shared in common was their experience of political contest and wariness in travel. Tensions between Celt and Englishman, rich and poor, and technocrat and libertarian were heightened through the struggle over the nation's highways. Science and technological expertise did not reconcile these differences but rather enabled political lobbies to pursue the development of the hinterland at the cost of other interests.

Infrastructure had become so vital to economic livelihood that every interest in the nation competed for it. That competition did not make a more social or peaceful nation. Rather, political struggles created a world where the rich policed the poor, where landowners smarted against the powers of the state, and where the economic success of a poor region like Scotland could collapse the moment its lobby for infrastructure weakened. The stakes of failure were daunting. As vagrancy laws were passed,

INTRODUCTION

radical bodies of travelers like the Methodist itinerants were broken up, their networks were dismantled, and their politics were tamed. When Scotland failed to secure money for its roads, the highways and bridges that had once been the best in Europe became ruins within a generation. Farmers could no longer afford to carry their grain to market, and the region that had once been emblematic of enlightenment plunged back into a preindustrial subsistence economy. The world of infrastructure was fraught with struggles for control where the very survival of radical politics or local economies depended on how players negotiated complicated laws and distant administrations.

In the world of infrastructure, letters and goods flowed on wagons and stagecoaches, connecting small islands and villages with great cities. The seemingly cosmopolitan parade of progress masked tensions over which interests controlled the roads and whose vision would shape the future. The story of infrastructure is not that of a glorious "transport revolution" where connection triumphs and ideas are never lost. Visions attributed to the road included the elimination of poverty, the ideal of participatory government, and the creation of radical political cultures, but those glowing futures foundered on the grim reality of who controlled infrastructure.

In a world where the flow of information is again being designed around hopes that connection will bring prosperity and participation for all, the story of the infrastructure state holds warnings. Infrastructure designed to enable the participation of poor people and poorer regions requires centralized control. Centralized control brings with it the specter of exclusion. The boards of civil engineers and other expert bureaucrats who oversaw road design were free to ignore criticism from fishermen and the victims of eminent domain. The British road system was the tool of different interests in turn, first the army and radical political travelers, then the poor hinterland, and eventually libertarian villagers. Under each regime, one part of the nation was connected while another was cut off, overcharged, or infringed on.

No people in the modern world can flourish without infrastructure, whether village farmers in the mountains, libertarians in wealthy villages, or radicals scattered across many towns. But only a government that reflects their interests equally can design a form of infrastructure that serves them all. Such a representative, participatory government, reflecting the interests of many constituents, was exactly what Britain failed to build. As a result, Britain's public infrastructure was crumbling by 1848, only half a century after its construction.

Here is a lesson for all who would end poverty or design their markets around the principle of participation. In the modern world, infrastructure is our principal tool for forging new communities, but it cannot outlast the control of governance by visionaries.

1

MILITARY CRAFT AND PARLIAMENTARY EXPERTISE

The Institutional Evolution of Road Making

> When work of importance is required, nobody cares to ask where the man who can do it best comes from, or what he has been, but what he is, and what he can do.
>
> —SAMUEL SMILES, *Lives of the Engineers*, 1861

Just before his death in 1834, bedridden and coughing in "bilous derangement," Thomas Telford chose to defer all interaction in favor of commemorating the technology responsible for his renown. A lifetime bachelor, now completely deaf and virtually reclusive, Telford was happy to enshroud himself in the finer specifications of paper weight, copper engraving, and typography, tools of political persuasion whose use he had refined during his entire professional life.[1] Between 1805 and 1811, Telford produced a remarkable corpus of documents that differed from anything earlier engineers had accomplished. These reports delineated in masterful detail the shapes of Scottish and Welsh mountain ranges. They showed in section and plan the routes of all contemporary paths, military roads, and wagon trails through those mountains, and they proposed new routes, more level and direct.[2] In 1834, Telford was still working on the same project of printing and persuasion, gathering the best visuals of his career for the richly illustrated appendix of the autobiography that would secure his reputation as the founder of civil engineering. Joining him was Edmund Turrell, the copperplate engraver, who on Telford's nomination had been elected to the Institution of Civil Engineers in 1826.[3]

The collaboration between Telford and Turrell reflected a transition in road making bound up with changes in the political sponsorship of road

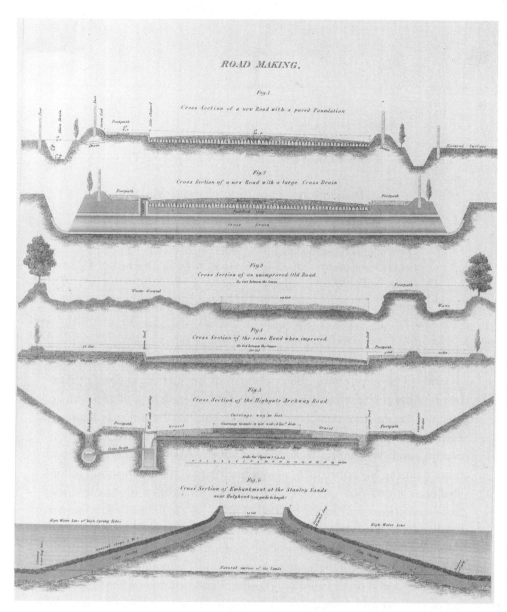

Thomas Telford and Edmund Turrell, "Road Making," in *The Life of Thomas Telford, Civil Engineer* (1838), from ideal road sections by Telford and Aaron Arrowsmith originally published in Coms. for Improvement of Road from London to Holyhead, and from London to Liverpool, *First Report*, 1824, appendix, p. 41. Copperplate engraving.

construction. Modern road construction emerged in the military laboratory of Scotland between 1726 and 1773 as a craft known to soldiers and surveyors. It evolved into the modern profession of civil engineering between 1803 and 1848, characterized by the practices of engraving, quantified evidence, and political lobbying, all of which justified road construction before the select committees that had begun to sponsor parliamentary road building in the name of national trade. Both stages of road making, the military and the parliamentary, set major precedents for the development of modern engineering. Military road builders developed new methods of sighting, foundation building, labor management, and trigonometric surveying, tools that enabled roads of greater distance, geographic penetration, and permanence than ever before. With the onset of parliamentary road building after 1800, a new breed of surveyor began to perform new functions focused not on better construction but on political persuasion and grounded in the amassing of quantitative facts about the potential benefit of collective roads. These parliamentary surveyors, or civil engineers, developed innovations that offered little in the way of better construction techniques.

Rather, the advances in trigonometry, mapping, and surveying after 1800 were developed to measure the mass of earthworks in the style already made. These measurements allowed engineers to quantify possible future costs and persuasively demonstrate their control and authority. Instruments and equations were both tools for persuading the readers of reports. As engineers gained greater trust through their use of hard numbers and lucid diagrams, the process of government endorsement for projects was streamlined, and engineers were able to ratify, fund, and execute projects of a scale, expense, and ambition that far surpassed those built by military regimes. Having added the skills of lobbyist and accountant to those of road builder, the civil engineer gained access to enormous sums of money that allowed the reconstruction of roads and landscapes across Britain. Britain's distant islands, wastelands, and difficult mountains were subsumed into a new kind of relationship with the center: the technological, trigonometric, measured connectivity of national space.

Remarkably, Thomas Telford, who did so much invention for canal and bridge making, invented practically nothing for the construction of the nine hundred miles of road he supervised for the British Parliament. There, he concentrated rather on matters of illustration and persuasion, working with men like Edmund Turrell to design instruments to quantify the length of potential roads and their probable cost, and diagrams to display that information better. Telford's intellect, like that of most civil engineers of his generation, was bent on designing new tactics for increasing support among the members of Parliament.

Turrell was one of an elect cadre of diagram makers who prepared Telford's drawings for parliamentary committees, and Turrell's coordination of information as an engraver was essential to the success of Telford's projects in Parliament.[4] In addition to his hundreds of unprecedented diagrams representing bridges, construction instruments, and road foundations in vivid detail, Turrell had perfected the form of the etching needle and drypoint and had published widely on the chemical process of engraving.[5] Thanks to the efforts of engravers like Turrell, parliamentary committees had been engrossed by ten-foot-long maps of contemporary and proposed roads, sections and diagrams illustrating how they would be assembled, and new trigonometric equations and extensive tables that documented the exact savings these projects were supposed to provide for the average traveler's journey.

These diagrams, more than road building, represented the real advance in technology. The "macadam" roads of gravel with their thick foundations were actually the invention of military engineers working in Scotland some fifty years earlier. Telford and Macadam merely built military-style roads on a broader scale than had ever been executed before, working with government to deploy military technology across local constituencies in England, Scotland, and Wales.

Bewitched by the cult of celebrity that enshrouded John Loudon Macadam and Thomas Telford by the end of their lifetimes, historians became accustomed to writing as if these engineers invented a far wider series of methods than they actually did.[6] But the century of road building was the work not of solitary genius, but rather of armies, parliamentary committees, politicians, soldiers, and parish surveyors.

Historians have used the story of inventors to characterize Britain's path to economic expansion in terms of the triumph of the free market. To date, when historians consider transport's role in the Industrial Revolution and creation of the nation, they describe impressive toll roads and canals, pioneered by entrepreneurs, and turnpikes connecting the capitals of industry.[7] Their tale presents an image of budding markets linked by private turnpikes, not of colonial capitals connected by parliamentary highways. In fact, both the construction and finance of the transport revolution were set in gear by government funding, made possible by the emergence of a state bureaucracy far earlier than historians have hitherto understood.[8] The military engineers who invented the construction tools of the transport revolution were deployed by the government, as were the parliamentary engineers who pioneered the shape of modern civil engineering. Turnpike trusts and local parishes built most of Britain's modern roads, but their tools reflected construction, surveying, and finance techniques invented at the level of government. Government, rather than private enterprise, supplied the great funds that drove technological innovation. The story of how the state drove the transport revolution refutes the myth that Britain rode to prosperity in the absence of government.

Military Building

The origins of military roads in Britain date to the 1720s and the expansion of empire that took British soldiers into the Scottish Highlands. From 1726 to 1745, England was engaged in policing the Scots clans who had threatened to install Charles Edward Stuart as a rival sovereign. Roads were essential to the circulation of English troops, and as late as 1745, no road thick enough to carry cannons crossed the Scottish border. Writing in his report to George I in 1725, General George Wade, the commanding officer, explained the "great disadvantage which regular troops are under, when they engage with those who inhabit mountainous situations." He compared the Highlands to the Savennes of France and Catalonia of Spain, all homes to traditionally free peoples who resisted attempts to subdue them. England's task would be formidable, he concluded. The only possible means of taming such fighters was to change the landscape itself. "Roads and bridges," he wrote, alone could render Scotland subject to English power.[9]

Begun in 1726 and built over the next eleven years, General Wade's roads linked the garrisons at Forts William and Augustus with Inverness, extending along Loch Ness and through to Dunkeld, Dalwhinnie, and Crieff. They totaled about 250 miles of road, averaging sixteen feet wide. After the second rebellion in 1745, another round of roads connected Dumbarton Castle, Stirling Castle, and Loch Lomond with the Bridge of Fruin and Duchlage. All the construction work was carried out by military labor, and soldiers continued to maintain the roads until 1815, when maintenance was transferred to a parliamentary committee. By this point almost 900 miles of roads had been built, created, and maintained by parliamentary grants of £4,000 to £7,000 a year, a burden that continued to haunt the military budget into the nineteenth century.[10] The roads were the first instance in Britain of a large-scale road-building project coordinated from above. In the process, the military began synthesizing the best knowledge available about land surveying, sighting, foundation making, and labor management.

The first set of tools concentrated on the practical and visual methods of sighting a route. Planning roads of territorial complexity meant laying new lines across varied landscapes such as those familiar to English surveyors from the flatlands to the south, but also through overlapping hills, around precipices and rivulets, and past the lines of property owners while still proceeding in as direct a fashion as possible. Edmund Burt, one of the surveyors employed in road building, took care to explain the problem of sighting mountain roads to those at home in England who had never attempted it. "Let us suppose," he wrote, "that where you are the Road is visible to you for a short Space, and is then broken off to the Sight by a Hollow or Winding among the Hills." English travelers had experience of watching the road disappear once or twice like this. In Scotland, however, the interruptions that staggered the line of sight were multiplied. After the first interruption, "the eye catches a small Part" of the road "on the Side of another Hill, and some again on the Ridge of it; in another Place, further off, the Road appears to run zigzag, in Angles, up a steep Declivity; in one Place, a short horizontal Line shows itself below." The bits of broken road seemed scattered over the entire landscape, mounting up and down in unpredictable patterns, until "the Marks of the Road seem to almost be even with the Clouds." It was disconcerting enough to see the linear stripe of the road mangled across the entire reach of the eye. Worse yet was the impression these changing patterns made on the moving traveler: "I need not tell you, that, as you pursue your Progress, the Scene changes to new Appearances."[11] Mountain routes, seemingly constantly changing their shape, frustrated the eyes of English travelers.

Military surveyors could make visual sense of the landscapes that confronted them only by applying newly discovered knowledge of vision and perspective to the task. The new landscapes appeared to eyes unaccustomed to them as a kind of visual riddle. Burt himself fumbled for words to convey to Englishmen the cipher that was the foreign landscape: "If I should pretend to give you a full Idea of it, I should put myself in the Place of one that has had a strange preposterous Dream." Surveying these passes,

Burt understood, was a work of mental effort, for the first response in viewing these new landscapes was visual confusion and psychological bewilderment: "To stop and take a general View of the Hills before you from an Eminence, in some Part where the Eye penetrates far within the void Spaces, the Roads would appear to you in a Kind of whimsical Disorder."[12] To the perplexed eye, new theories of vision and perspective offered helpful tools. Contemporary theories of perspective suggested the basic outline of the vanishing point and how overlapping planes faded into the distance.[13] From these theories, military surveyors extrapolated their own explanations of what they were seeing. Burt asked his readers to consider how it was possible still to look at the road when it was so far above them: "How can you see any Part of the flat Roof of a Building, when you are below?" Burt was borrowing an analogy for explaining the disappearance of mountain roads from the contemporary French philosopher Jean Dubreuil, who had explained in his *Practice of Perspective* how a "mountain may have its Top above the Horizon," and that for objects above the horizon "the Top" becomes "invisible."[14] Burt elaborated Dubreuil's theory of perspective and used elements of John Locke's theory of vision to describe how mind and eye wove together bits of landscape in seemingly arbitrary patterns through the workings of perspective: "The Eye catches one Part of the Road here, and another there, in different Lengths and Positions; and, according to their Distance, they are diminished and rendered fainter and fainter . . . till they are entirely lost to sight."[15] Recent treatises on the eye and its relation to perspective made the complex patterns of landscape in Scotland less frightening and gave surveyors confidence in their sight sufficient to keep them experimenting with the process of sighting of which their work consisted.

These visual difficulties proved to be technological problems in the task of making roads: sighting the straightest path of a road past overlapping peaks in perspective, despite cutbacks and natural barriers, posed difficulties utterly new to the English science of road building. Sighting had to be reinvented and reimagined for the purpose of making new routes. To figure out where a straight path should go, surveyors developed various tricks. Large boulders were used as "Rules" or landmarks for sighting straight lines, which were followed "as far as the Way would permit," tracing the path most likely "to shorten the Passenger's Journey."[16] Working in short stages, the rule stones broke down the elaborate pattern of lines vanishing in perspective. They made it possible for surveyors to conceive of the entire project of making a way across the mountains by combining smaller projects of limited scope, where the surveyors could reckon the best route by sight.

New techniques for measuring and delineating the landscape—surveying and cartography—also appeared first during the military conquest of Scotland. Used first for the purposes of reconnaissance, these techniques eventually allowed road engineers to project the most even route for a new highway. Between 1747 and 1755, England undertook the military survey of Scotland, the first trigonometric survey conducted on British soil. The chief surveyor was William Roy, who would become a major collector of scientific instruments and the driving force behind the Trigonometric Survey, later the Ordnance Survey of Great Britain, established in 1791.[17] The military survey of Scotland put to the first organized, widespread, and creative use the talents of a profession that had its own place in civilian society: the surveyor. Military surveying, never before applied to British landscapes, remained in its infancy in Britain, lagging behind the mobilized mapmakers of territorial military powers like France. The surveying project, directed by William Roy and David Watson, brought together an elite cadre of officers and landscape surveyors who were attempting the most ambitious surveying project yet executed on British soil. Eventually, accurate maps would be put to use for plotting potential routes.

Military advances in surveying represented a remarkable expansion over contemporary civilian methods. Adam Martindale observed in 1702 that "the Country aboundeth with such as, by their Inclination and Interest, are prevail'd with to take Pains in measuring Land, that for want of better Instruction, use ill divided Chains, and tedious Methods of Computation, which makes their Work intolerable troublesome, if Exactness be requir'd."[18] The training of most surveyors focused not on measurement with chain and compass, but on the accurate delineation of shapes on paper. An entire chapter of Edward Laurence's book *The Young Surveyor's Guide* (1716) was dedicated to "an useful collection of all transparent Colours proper for Beautifying Maps, Charts, &c."[19] Describing the tool set of a professional surveyor, George Adams listed "two drawing pens, and a pointrel" (for engraving) and "a small set of Reeves's water colours" (for applying pigment to a map).[20] Henry Wilson, writing in 1743, explained the surveyor's talents as a function of the map's "lustre and beauty," which a skillful artist would render by marking "black lines" around each "field and inclosure" with a "small Pencil and some transparent Colour." He explained the shadowing of "mountainous and uneven Grounds of Hills and Valleys" and the "lively Colours" great cartographers used on "Bogs, Groves, Highways, Rivers, &c." The accomplished cartographer was, in essence, another kind of artisan, crafting an ornament of beauty: "These things being well performed, your plot will be a neat Ornament for the Lord of the Manour to hang in his Study, or other private Place; so that, at pleasure, he may see his Land before him, and the Quantity of every Parcel thereof, without any farther Trouble."[21] Although most such surveying took place for ornamental or legal purposes rather than ones of projecting possible routes, early eighteenth-century cartography was characterized by artisanal skills of close looking and craftsmanlike rendering.[22] The colorful surveys that were produced simplified the task of making sense of territory.

This artisanal process of observing and transcribing visible landscapes in cartography was not entirely distinct from the allied art of sketching scenery. The brothers Paul and Thomas Sandby, later distinguished for watercolor and architecture, were both employed by the military in Scotland after 1730.[23] There they joined half a dozen other artistic civilians, recruited by the Board of Ordnance and trained by the French military surveyor, Clement Ladurie, at the Board of Ordnance's classrooms in the Tower of London.[24] Ladurie exposed the artists to recent French treatises, like Nicolas Bion's *Construction and Principal Uses of Mathematical Instruments*, first translated in 1723, which related the principles of the compass and protractor, recently developed on behalf of the French army.[25] They also learned about the accurate use of the chain to ascertain distances and about measuring angles with a theodolite or compass, practices set down in English treatises only in the 1720s and 1730s. The Ordnance Survey brought the first British surveyors into contact with advanced French mathematics to quantify and render the landscape into geometrically verifiable maps, sealed with theodolite and compass.[26]

But even these first mathematical surveyors depended on artistic sketching. An artist's rendering of enemy encampments was one of the standard tools used for military reconnaissance. In the papers of Paul Sandby, several watercolors of rebel encampments describe his work as a scout. Other watercolors show newly constructed bridges or passes awaiting bridge construction. The Ordnance Survey's artists-turned-trigonometricians toured Scotland under General Wade and then joined the military survey of Scotland under Lieutenant-Colonel David Watson. As part of their tour as surveyors, they made a series of sketches of the landscape that served a dual purpose for military surveying. First, visual studies of the Scottish landscape perfected their ability to discern patterns in the complicated, overlapping hill country that so confused Burt. Second, these studies supplemented cartography in documenting the landscape for intelligence purposes. Sketching and watercolor, then, were as much a part of the surveyor's tool set as measuring angles and distances, and they remained so well into the nineteenth century, when teams armed with theodolites quantified only major boundaries like waterways, coasts, and roads, while the topography was sketched in by eye.[27]

Advances in trigonometry continued to receive their greatest support from the military, including several of the individuals who had been involved in the road-building projects in Scotland. William Roy, who had directed the military survey of Scotland, became England's surveyor of coasts and in 1785 directed the first steps of the Ordnance Survey.[28] Through the work of Roy's surveyors (among whom were the Sandby brothers), the advanced methods of surveying carried out in the Highlands were brought back to the metropolis, where they received continuing government patronage. Indeed, of the dozen surveying manuals published between 1770 and 1790, four were published by individuals with some relationship to Roy and the Ordnance Survey and were indeed dedicated to the survey's directors. George Adams, for instance, dedicated his 1791 treatise on surveying to Charles, Duke of Richmond and Lennox, the master general of the ordnance. Military patronage of surveying encouraged the redaction of surveying knowledge and the publication of trigonometric manuals for a wider audience.

Through the encouragement of military practice, a wide variety of tools became available to surveyors intent on more clearly reckoning the lines of geography. Triangular, elliptical, and proportional compasses—useful for laying down places on a chart, measuring angles, drawing ellipses, and calculating proportions—came into wider circulation in the 1770s.[29] Parallel rules, protractors, and newer, standardized theodolites also were more widely introduced. Advanced mathematical texts began to appear that influenced later works on surveying. Benjamin Donn, a county surveyor in Devon who lectured on mathematics in Bristol, popularized trigonometry through his *Mathematical Essays* (1764) and *Essay on Mathematical Geometry* (1796). The geometric mechanical methods of describing circles were laid out for the first time in Joseph Priestley's *Perspective*

(1770). Together, these mathematical avenues into geometric description helped surveyors specify points and distances more accurately.[30] They were brought together as a whole by a new method of plotting points, not freehand, in relationship to one another, but on a regularized grid. This "new method of plotting" on a grid, developed by Mr. Gale, was touted by surveying textbooks as a godsend: "This method is much more accurate than that in common use, because any small inaccuracy, that might happen in laying down one line, is *naturally corrected* in the next; whereas in the common method of plotting by scale and protractor, any inaccuracy in a former line is *naturally communicated* to all the succeeding lines."[31]

Eighteenth-century geometry was first and primarily furthered by the military surveyors who participated in the military survey of Scotland and who afterward taught at the newly instituted Woolwich Academy for military officers.[32] Gradually, military trigonometric practice was translated for the common county surveyor of properties. George Adams's 1791 essay on trigonometric surveying translated military knowledge for the domestic surveyor and engineer, introducing the "first principles" of trigonometry, the design and use of mathematical drawing instruments, and a series of geometric problems for practice.[33]

Parish road surveyors were quick to capitalize on the new technology with the help of new surveying manuals that translated mathematical theory for their benefit and showed the direct and clear application of advanced trigonometry to their practical and local needs. William Gardiner's 1737 treatise *Practical Surveying Improv'd* dedicated a full chapter to the surveying of counties, explaining the surveying of public ways as the surveyor's primary duty. Drawing on the early trigonometry of Nikolas Bion, Venterus Mandey, and Edmund Stone Cunn, which had influenced the military projects, Gardiner offered a step-by-step guide to taking bearings with a theodolite, measuring distances with a circumferenter, and measuring the angles between the road and every landmark along the way, including "all churches, wind-mills, great houses, &c." Gardiner explained the everyday tasks of the surveyor walking the countryside: "As I measure on in the Road, I note every Road and Lane that turns out of it; and whether it is on the right hand, or the left, and to what place it leads; and whether it inclines forward, or backward, or is nearly at right angles." Every time he noted a new alley, lane, or property boundary, he made "marks" in his book, and for each road he took an additional set of calculations: "I plant my Theodolite, and take the Bearing, and measure to POINT 1 in it,

MILITARY CRAFT AND PARLIAMENTARY EXPERTISE

and mark both places to be found again." The surveyor was not only the documenter but also the judge of these roads. Gardiner explained, "I write the quality of the Road, whether good or bad; and where hilly." His aim was to document as exactly as possible: "And thus the Road will be agreeably fill'd with remarks."[34] A satisfying map was a map full of information.

Military standards for trigonometry and record keeping spread dynamically through the trusts and parishes. A constantly increasing tide of improvement generated a broadening stream of trained surveyors in most rural counties, as Peter Eden's evidence demonstrates for rural Norfolk.[35] By the end of the eighteenth century, schools taught surveying, and more than 2,500 villages had a resident surveyor; the number of new surveyors per year rose from 17 in the seventeenth century to 119 for the period 1784–1850.[36] Surveyors were basically unknown in Scotland and Ireland in the seventeenth century—there were only 10 surveyors in Scotland in 1731. Their numbers exploded in the 1750s and 1760s with the reorganization of forfeited estates and the creation of planned towns.[37] With the canal boom of the 1760s and 1790s, the numbers of civil surveyors throughout Britain multiplied as they worked alongside engineers and architects; cartographers and land surveyors were members of the Society of Civil Engineers founded in London in 1772.[38] By the 1770s, ordinary surveyors' manuals were adopting the language, tools, and theory of scientific trigonometry recommended by Gardiner and his followers. Thomas Breaks, in his *Complete System of Land-Surveying* (1771), gave detailed instructions for county surveys of roads. He suggested that the surveyor proceed by checking the basic survey of a landscape by using the theodolite and perambulator (or odolite) to take the distance and angles between each "memorable Monument within the County," thus double-checking these shapes against one another.[39] By the 1790s, the basics of trigonometry and land surveying, refined into basic principles, were being taught to the pupils in boys' schools from textbooks like *A Treatise on Land-Surveying*, written by Thomas Dix, a teacher at Oundle.[40] Parish surveyors quickly adopted these tools, eager for further help in their duties as road managers and witnesses in property disputes.[41] The striking spread of information about surveying prompted the gradual expansion of the surveyors' profession and wider acquaintance with military knowledge of the landscape.

Once territory had been accurately surveyed, it was possible to plan routes in an entirely new fashion: the possibility arose of plotting a prospective road on a premade map of the territory rather than engaging in road sighting as the laborers moved across the landscape. Where the ant's eye of sighting took in only the nearest possible vantages, the bird's eye of plotting made possible an actual determination of the most even route. The routes that followed ancient trackways gradually began to yield to direct paths. In Lancaster, the Durham Road to Shotley Bridge, which originally followed the "circuitous" high ridgeway, was replaced in 1810, when the turnpike trust was able to negotiate a path that led "directly through the vale."[42] By the 1800s, military advances in surveying technology had opened up a new set of possibilities, and surveyors could point knowingly to places where alterations might be made.

Maps inspired even grander ideas among surveyors of vision. In 1766, John Gwynn included half a dozen surveys of proposed routes to rationalize London's streets with his proposals for improvements to the transportation of the city and its surroundings.[43] John Holt, a surveyor in Lancaster, recommended in 1795 that all turnpike acts be required to submit "an exact plan of the proposed roads" whenever they asked for parliamentary approval. He pointed to the fact that such a survey of the road's "connection with the neighbouring towns" and "an accurate *section* of the whole line of road" would allow the parliamentary committees involved to "judge of the propriety of future applications for making new or *amending old* turnpike acts."[44] Equipped with the military tool of the bird's-eye view, surveyors could speculate about realigning entire systems of roadways.

The success with which route planners deployed trigonometry depended largely on the organization of the local government. Arthur Young found that the Salisbury to Winchester Turnpike was "broad enough for three carriages to pass each other" and followed "straight lines with an even edge of grass the whole way," "more the appearance of an elegant gravel walk, than of an high road."[45] Lazier or cheaper trusts might simply follow the boundaries of the properties of individuals who were most easily reconciled to the project. The Barking Road in London, for instance, the result of a turnpike trust performing construction in 1812, according to a perturbed visitor, "bends at every group of houses, into the direction of the next group, lying in any thing like the proper direction. This is remarkably the case east of Rainham, twelve miles from London, beyond which place the road is a complete series of zigzags."[46] In London, the ground was flat; the zags in these roads were the result not of boulders and precipices but of householders unwilling to sell at the trust's price. Regardless of the skills of the individual surveyor, variations in the cooperation of

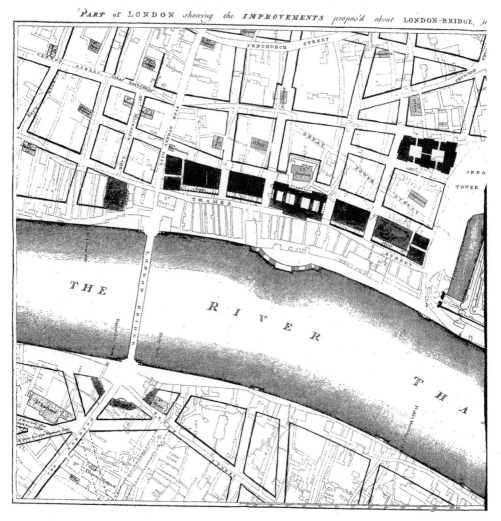

John Gwynn, "Part of London Shewing the Improvements Propos'd," *London and Westminster Improved* (1766), p. 155.

turnpike trusts, local government, and local landholders could produce great deviations in the route.

The personal experience of the road surveyor chosen accounted for a great degree of difference in the success of different road projects. Individual experience determined whether a given surveyor had exposure to military methods. Most road builders determined the path of new roads by a mixture of guesswork, negotiation, and apathy. New roads being routed over wastes or commons could take whatever path the surveyor deemed most direct and least expensive. This was rarely a scientific calculation. The turnpike surveyor hired for the Marsden and Standish Road determined a path for the highway that "took it over deep marshes." When John Metcalf was invited as the surveyor, he reckoned that the cost of draining and foundation building in the marshes would outweigh the expense of routing the road another way, and by providing some estimated figures, he persuaded the turnpike trust to lay the road by a more circuitous but drier way instead.[47] The tools of military surveyors—adept at reckoning the boundaries of property, the best path for a road, and the number of householders involved in a decision—were generally beyond the reach of most road builders. As Mr. Mew explained of his friend John Metcalf: "The plans which he makes, and the estimates he prepares, are done in a method peculiar to himself, and of which he cannot well convey the meaning to others."[48] Metcalf's individual talents included an innate ability to calculate figures in his head, a skill difficult to learn by mere imitation. Textbooks for teaching trigonometry were, as we have seen, in their infancy. In the era of military road building, trigonometric knowledge was barely codified as a set of tools easily conveyed from one practitioner to another. Rather, individual surveyors transmitted the standards of available knowledge on a personal basis, and individual surveying habits varied widely as a result of slow dissemination.

The surveyor's individual personality provided another essential component of successful road building. Wit, political experience, and social negotiation, rather than direct calculation, played the main roles here. John Metcalf was lucky to be gifted at the practical skills of negotiation, an essential tool for the surveyor who wished to straighten roads that abutted other properties whose shape prevented a direct route. One of Metcalf's roads around Harrogate passed through an area that was the property of an old woman, whose house held a "common right upon the Forest." Metcalf negotiated with her and bought the house and property, cut the road directly through the land, and resold the house afterward at a profit of two hundred pounds.[49] A good road builder with Metcalf's experience as a salesman could make these personal negotiations, but they were the job of the road surveyor acting alone, without lawyers, accountants, or other brokers, and operating beyond even the support or knowledge of the trust. Rather than face such negotiations, most surveyors were content to run their roads in the direction that required the least negotiation. Such skills depended on personal acumen, experience, and charisma, features not easily taught. A century later, the historian Samuel Smiles meditated on Metcalf's success as a result of his "adventurous career" in the army and his blindness, which required a replacement of the study of books with the study of men. Smiles reflected that Metcalf "could read characters with wonderful quickness, rapidly taking stock, as he called it, of those with whom he came in contact."[50] He hardly offered a recipe for the many parish surveyors of Britain, who did not possess the means of reduplicating either Metcalf's personality or his skills at negotiation. The success of each turnpike and parish in improving roads thus reflected largely the adventures, experience, and personality of the individual surveyor it had on hand.

Military innovations in sighting and plotting enhanced the range of tools local surveyors could use to find the best route through which to run a road. Dramatic deployment of both sighting and plotting characterized some developments, like Metcalf's roads and Gwynn's plans. These innovations were not absorbed equally everywhere. The directness of the path of Britain's roads varied enormously, reflecting differences in geography, local government, and personality. Military technology offered a recipe for perfect roads, but technology spread only as quickly as human institutions allowed.

After surveying the road and erecting landmarks, the next stage of road making was clearing and building. The length of the roads and the varied territory through which they ran required builders to experiment with new approaches to construction. The routes passed through "moors, Bogs, rugged, rapid Fords, Declivities of Hills, entangling Woods, and giddy Precipices." Building on these terrains was, as Burt put it, an "extraordinary Work." To remove boulders from the path in the moors required tunneling alongside and underneath the boulder and so turning it deeper into the marsh, where it could not protrude into the path of the pavement. Thick-grown forests posed a different challenge that the builders had exerted great pains to meet: "The Trees, for the necessary Space, have been cut down and grubbed up; their fibrous Roots, that ran about the Surface, destroyed; the boggy Part removed; the Rock smoothed, and the Crannies firmly filled up." On steep mountains, the builders used winding cutbacks, "supported on the Outside of the Road by Stone Walls, from ten to fifteen Feet in Height." In bogs and mosses, areas of spongy earth, the roads required an elaborate foundation of knotted bundles of heather that floated the gravel stacked over them. These strategies for filling in foundations were new and effective. They would form the basis for road building for the next century, and even in the 1820s John Loudon Macadam would become famous for disseminating these techniques on a wider scale. In Scotland, the result of these labors was a path, Burt wrote, "as smooth as Constitution-Hill," where one could gallop "for Miles together in great Tranquility."[51] The techniques for making such a path—sighting, boulder clearing, and foundation building—were new tools, deployed with enormous funding and impressive discipline, ambitiously conceiving of road surveying anew in both theory and practice.

Through a slow trickle of imitation and publication, the advanced surveying methods of the military gradually found their way into the regular tools of parish surveyors. A single individual was responsible for disseminating most of these methods in England. Blind since birth, John Metcalf had traveled as a soldier with Wade's troops over most of the Scottish roads. Upon his return to England, he became a private contractor and surveyor, a prolific force for spreading the practice of military foundation making across the nation. According to Mr. Mew of Lancashire, a personal acquaintance of Metcalf, "Most of the roads over the Peak in Derbyshire have been altered by his directions." Smiles enumerated his roads, which stretched over a total of 180 miles and earned him a payment of £65,000, constructed in Lancashire, Yorkshire, Derbyshire, Cheshire, and Staffordshire in a career that stretched from 1765 to 1792.[52] Metcalf was responsible for generalizing improvements in foundation laying, a mixture of close attention to materials and the experience of Scottish highway builders. Henry Sacheverell Homer described Metcalf's method for laying a foundation, which by the 1760s became standard throughout Britain: "Pavements constructed with common Pebbles, which are generally of a very hard Substance, and if well bedded in Sand, rightly disposed, sufficiently rammed and well covered."[53] Like the military road builders, he plowed roads through virgin forest, using heavy equipment to carve a fresh path. For the road between Harrogate and Harewood Bridge, "he had a wheel-plough drawn by nine horses through the forest, as the best and most expeditious way to get up the roots of whin and ling, in parts where they were strong." He also developed methods of road building across bogs not unlike those of the military in Scotland. For his road through the Pule and Standish common, where boggy water made the ground so weak that horses could not deliver stones there, Metcalf adopted the military system of sinking deep foundations and filling them with the available materials.

He "ordered his men to bind heather, or ling, in round bundles, and directed them to lay it on the intended road, by placing the bundles in squares of four, and laying another upon each square, pressing them well down."[54] On top of this springy, heather foundation, Metcalf had the men layer stone and gravel, safe and dry above the moist earth below. By floating the road above the surface of the bog, Metcalf developed a means of building roads across even the soggiest territory. Carried south by Metcalf, military methods became, by 1760, the standard for British foundation making as a whole.

Metcalf also invented new strategies for leveling hilly territory unfavorable to road building. Before he built the road, he first set about leveling it "by building up the hollows, and lowering the hills."[55] Metcalf could create level roads where none had been before. In Huddersfield and Halifax, "Many were of the opinion that it was impossible to make a road over that ground." Metcalf's insights again derived from his experience in Scotland. Compared with those dramatic territories, the worst areas of the north of England were tame. The area around Saltershebble, for instance, was "very rocky," "full of hollows," and "boggy and rough," but lacked the extended stretches of peat bogs and mountain precipices surveyors had dealt with in Scotland. Metcalf used his career on these gentle hills to refine the available tools for creating level surfaces: arches and filling in, and blasting and carting out. On the road from Dock-Land Head to Ashton-under-Lyne, "He raised one hollow nine yards, and built sufficiently on each side to keep up the earth, with battlements on the top; for which he received two thousand pounds." Where hills got in the way, they could be blasted and then carted out, as on the Whaley and Buxton road in Derbyshire, where the proposed road "went over a tedious piece of ground called Peeling Moss; the whole road being four miles in length, with some part strong rock, which was to be blasted with gunpowder." This work of haulage was difficult and expensive in terms of horseflesh, but Metcalf was undeterred: "Though he lost twenty horses in one winter, he was not discouraged; observing that 'horse leather had been dear a long time, but he hoped now to reduce the price.'"[56] Blending military tools, innovations, determination, and organization, Metcalf's roads conquered new territories for passage.

These schemes caught on. In the 1770s, one surveyor advocated the thick, gravel foundations of the military-style gravel road, explaining how "minute Particles of different Substances" "unite closer (and as Road Makers term it) bind harder" through the force of their inherent "Powers of Attraction."[57] The use of broken stone spread so that Joseph Hodgkinsen advised in 1794 that bad roads might be made better were the surveyors only to "break the stones small and of equal sizes . . . that no water may lodge" and that they "cement together" into a "good and durable road."[58] Boggy roads lifted on floating foundations of heather were reportedly still being made anew in Scotland and were recommended to further areas by the Board of Agriculture in 1794.[59] Cuts and fills were immediately adapted to canals and river navigation designs, where leveling was even more essential than for roads.[60] They caught on more slowly for roads. The military innovations proved influential. Writing in 1767, Henry Sacheverell Homer praised the military-style work that Metcalf had accomplished and his influence on the entire profession. Metcalf's arts would have been "chimerical" to "common Surveyors" of the past with their "contracted Ideas." Metcalf's military arts of "raising Vallies and sinking Hills" were gradually changing "Custom" and enlarging "their conceptions," familiarizing them with "a Conviction both of the Practicability and Utility of such Schemes."[61] By the 1770s, the anonymous author of *Schemes Submitted to the Consideration of the Publick* was recommending "lowering the Hills" and "filling up the Valleys" of the roads between London and St. Albans, "where the road is almost a continual Winding, and the greatest Part between Barnet and St. Alban's a continual Hill and Dale."[62] Soon thereafter, Archibald Campbell, a Scottish surveyor who had studied the military roads, made similar advances in cutting a road around the base of the mountains that separated Kintyre from the rest of the country.[63] Between 1740 and 1790, reports of voluntary road leavening by individuals, parishes, and turnpikes were arriving from Nottinghamshire, Yorkshire, and beyond.[64] Indeed, Metcalf's blend of military foundation making with advanced leveling became the rule by which later civil engineers were building well into the nineteenth century.[65]

Despite the gradual dissemination of military technology, many local roads displayed variations in construction, management, and maintenance beneath the standard established by military road building. In his numerous tours across the nation, Arthur Young documented vast differences, reviewing the major turnpikes in the country in conditions ranging from "good" and "middling" down to "a line of vile deep rutts cut into the clay," "a disgrace to the whole country."[66] As W. A. Provis, one of Telford's assistant engineers, testified, the Welsh part of the Holyhead Road had been paved before improvement "in an irregular and promiscuous manner, some of the stones standing near a foot above others, and in some places holes were left without any stones."[67] Despite the influence of military trigonometry, surveying and road building remained professions whose success was a function of local conditions and personality.

Local soil influenced much of the variation in road conditions visible across the country. Local roads had always varied with local access to materials. In 1727, John Houghton reported "that at *Deptford* they laid flints a foot thick, and then covered them with gravel, which makes a good road." Near Oxford, he found "a substantial high road, covered with chalk," and approvingly noted that "here cou'd lie no water except in the holes made with horse-feet and wheels, and I doubt not but those are easily mended."[68] Similar variations were still the rule when Arthur Young made his tours in the 1770s.[69] Although Houghton was impressed by the geographically specific solutions for improving roads, Young advocated a single, military rule for road improvement, applicable to all surfaces. He repudiated the folklore of "natural roads" that held that certain regions, such as Norfolk, were founded on such organically dry and firm soil that ancient trackways through them sufficed without any sort of management whatsoever. Young sniffed, "I know not one mile of excellent road in the whole county."[70] The provision of a single standard for improved roads meant that canny travelers like Young were more critical of varied, local solutions for road improvement. Where personality, local government, and soil conditions had once hindered the dissemination of advanced building ideas across the nation, Britons increasingly measured all roads against the single standard set by military improvements.

Finally, road building on such a scale demanded labor organization of new proportions. For both the erection of sighting boulders and the actual construction of roads from gravel across the moors, soldiers were involved in a primitive and time-consuming process of stone breaking and boulder moving.[71]

Stone breaking and road making were grueling tasks, and the soldiers responsible for them received sixpence above their normal pay in recognition of the hard labor they endured. Generals presented these added costs, like the costs of the roads themselves, to Parliament. Wade compared the discipline of his own troops with the plundering and pillage of German soldiers whose meager pay was supplemented by raiding the countryside, warning Parliament that peace would follow invasion only if the soldiers doing labor in Scotland were well paid and well governed. Sir William Wyndam speculated that that the Sinking Fund would easily handle the necessary finances.[72] On the basis of convincing, personal testimony from the generals, finances were procured.

So many men performing such hard labor in the remote wilderness could be effective only with a strong system of hierarchy in place. The five hundred soldiers who worked under General Wade were divided into "Parties of Men" organized into "Detachments from the Regiments and Highland Companies," each under the direction of a subaltern referred to as the "baggage-master" or "inspector of roads in North Britain." Over each of these was another set of officials. Noncommissioned officers directed the path of the road, working as "Overseers of the Works."[73] Exacting, hierarchical structure proved an efficient and effective means of coordinating hundreds of men across vast stretches of territory.

Parish road making gradually adopted these military strategies for managing laborers. Metcalf instituted new kinds of organization to keep his men at work that resembled the military in deployment, commitment, and rapid labor. Like the military, Metcalf divided his men into well-equipped teams: "He set men to work in different parts, with horses and carts to each company." He also set up encampments to keep the men nearer the labor, thus cutting the time for transportation: "He therefore provided deal boards, and erected a temporary house at the pit, took a dozen horses to the place, fixed racks and mangers; and hired a house for his men at Minskip, which was distant about three-quarters of a mile. He often walked from Knaresborough in the morning, with four or five stone of meat on his shoulders, and joined his men by six-o'clock."[74] Thus he completed the work in record time. He similarly rented inns along the way for his men in future projects. Other parish surveyors followed Metcalf's pattern. In Somerset, William Maton observed "labourers" who broke stone "with an one-handed hammer . . . in the midst of the rubble," following the military recipe for labor-intensive gravel making.[75] By the 1790s, Joseph Hodgkinsen was advising surveyors to "assign to each labourer a certain district to look after, one mile, half a mile, or more or less," managing each so that "each man will take a pride in keeping his district better or equal to his fellow-labourers on the same road."[76] Road surveyors also adopted the military style of highly organized labor, delegated in bursts of "3 or 4 days" at a time, during which Gardiner traveled with a band of assistants. Gardiner himself took the measurements and wrote down "every thing in a plain manner, and with all possible caution not to make any mistake," while his assistants drove down "a short stake in the place of every object left," marking the entire landscape that the surveyor had covered. The next stop, plotting these measurements on a chart, was done by Gardiner himself. Together, the surveyor and his assistants could cover the entire landscape. The gradual dissemination of military-style organization took the form of efficiently coordinated gangs of laborers that began to appear on road improvements across the country.

Parliamentary Roads

Parliamentary road building after 1803 deployed the construction techniques of the British military to connect London with Dublin and Edinburgh by projects of a scale entirely different from military ones. Parliamentary roads, unlike military ones, were permanent structures designed to carry the nation's commercial intercourse for decades and centuries to come. Military roads had been intended only to carry soldiers to police the Highlands for a limited time until the hostilities died down. Even then, the difficulty of travel was less important for the relatively few trips in any direction a regiment and its supplies might take, in contrast to the daily flock of comers and goers on most commercial roads. On the basis of the roads' permanence, road builders could point to the relative merit of expensive bridges and embankments that would ease the way for many travelers.

Parliamentary roads had a different relationship to funding than military ones. First, they were more expensive. Their wide roads and bridges, designed to ease the experience of travelers permanently, entailed greater expenditure per mile than the narrow foundations of the equivalent military roads. Second, because they were public ventures ratified by Parliament, their expenses had to be specified and approved in advance of a grant approved first by a parliamentary committee, then in a bill voted on by both houses, and finally by the Treasury. Pleas for funds for parliamentary projects therefore had to be ironclad, persuasive reckonings in advance, unlike military requirements, handled by the Board of Ordnance and then the Treasury, which could be flexible to meet ongoing needs.

Parliamentary building thus shifted the burden on surveyors from that of creators of roads and reconnaissance to that of projectors of possible roads and estimators of funds. Parliamentary engineers became inventors of maps, plans, statistics, arguments, trigonometric calculations, and measuring instruments that defined the role of a new kind of engineer, the civil engineer,

MILITARY CRAFT AND PARLIAMENTARY EXPERTISE

whose calling was related to the scope of national and permanent rather than military or local projects. Financial engineering made apprehensible the financial layout of potential projects of a scale never before attempted and thus made possible a new scale and permanence of building.

Between 1806 and 1835, a series of parliamentary committees collected information from expert witnesses about the best organization of national rules to govern the wheels of carriages and the administration of turnpike rules. These laws were nothing new. Loose regulations on stagecoaches had existed since the seventeenth century. Rules that had governed individual users of carriages since the 1760s, setting out requirements for axle length, the number of horses, the removal of trees from the road, or the width of wheels, could potentially improve the overall condition of road surfaces without requiring much administrative or bureaucratic overhead.[77] Carriers and cart drivers were required to paint their name and address on every cart, and their names were recorded in the books of the local justices of the peace to make it easier to hold them responsible for any infraction in the number of horses or thinness of their wheels.[78] Parliament had stipulated the public responsibility of turnpike trusts in each of the some one hundred turnpike acts passed in the eighteenth century. Acts of the 1780s and 1790s governed the number of persons a stagecoach could carry.[79] The 1806 parliamentary committee was instructed to review and clarify the many acts that governed turnpike roads and parish highways. It was charged with reviewing "what shape" of wheel was "best calculated for ease of draught and the preservation of roads," such that the hazy regulations about broad wheels could be clarified to keep up with new trends in stagecoach making in the 1790s and early 1800s.[80] This long tradition of government regulation of vehicles set the precedent for parliamentary collection of information about road traffic.

The committee was charged with suggesting ways to better enforce the existing laws, which were subjects of notorious evasion. In the 1790s, the Post Office had briefly tried to speed the delivery of mails by writing to local surveyors of highways where roads were bad. If the surveyors did not respond, the Post Office attempted to indict the parish by finding a judge who would rule against the surveyor. But indictments were notoriously difficult to enforce because there was no central body in charge of following through on whether the parish had been held to task after a ruling. Parliament's early attempts at regulation, which centered on the behavior of stagecoach drivers, met similar setbacks. Regulating stagecoaches was notoriously difficult because of the mobile nature of the offenders: summonses served on stagecoach proprietors were often in the hands of authorities of parishes through which the overburdened coaches passed, who could rarely afford to deliver the summons to the residence of the offender at a great distance.[81] Because of the lack of mechanisms that ensured the liability of offenders, legislative attempts to regulate the road system floundered.

To these insufficient solutions, nineteenth-century parliamentary committees proposed a radical answer: a centralized system of regulation that would consider traffic, commerce, pavement, and vehicles as a whole, including the best means of monitoring and enforcing the responsibilities of parishes and individuals. This was a titanic effort of hundreds of minds, requiring the collection of local practices from parishes across the three kingdoms, as vast a system as the writing of the French *Encyclopédie*. Indeed, the gathering of information promised to offer the basis for consensus.

Upon the fabric of expert information was built a new foundation for systematic policy-making at the national level. Looking back over differences in voting and taxation, the student of eighteenth-century government would witness crisis after crisis resolved by the manufacture of consensus. Indeed, the practices of collecting expert opinion had evolved in order to ensure consensus in the courts. Where testimony became so complicated that the judge's ruling smacked of interest, seventeenth-century judges had begun to call on local experts to testify. Experts again appeared in the 1780s to balance Treasury reports that were accused of party bias in the era of the civil list, and by 1787 Parliament obliged the Treasury to publish expert-overseen annual reports as a routine measure.[82] By the 1790s, the collection of diverse, quantitative information about the best form of regulation had at last become standard procedure for committees charged with the nation's laws.

On the grounds of expert testimony, the 1806 committee could offer a systematic approach that extended well beyond the regulation of wheel shape and stagecoach passengers. The committee proceeded by collection of information that it presented in the form of a sixty-two-page report of "useful suggestions" to be refined into a "System regarding the Public Roads."[83] The display of exhaustive, systematized, quantitative evidence had particular power in these years, when the sources of evidence accepted by the select committees were expanding and being refined. The result was presented under a grand title, "The Best Means of Preserving the Turnpike Roads and Highways of the Kingdom." It was the first overview of trade, pavement, and vehicles considered as a system.

The information printed in the minutes of evidence for the parliamentary reports was gathered widely. Parliamentary funding set up a system where self-styled engineers could possibly gain recompense and fame for better innovations. Macadam was voted some £2,000 for consulting with the Post Office on the best kind of road surfaces that would ensure speedy postal delivery around the nation.[84] John MacNeill was paid £527 for the purchase of his road indicator, which allowed surveyors to compare the friction of different surfaces.[85] Advising parliamentary committees was a well-remunerated job. Macadam and Telford were therefore joined by a host of rival, self-taught engineers who also emerged from the vast ranks of professional surveyors, architects, county magistrates, and instrument makers to offer new solutions to an emerging economy of infrastructure design. A new breed of engineer emerged out of this fierce competition for Parliament's attention and the possibility of pay. He was not strictly speaking a scientist in any sense, discerning natural laws or motivated by the quest for knowledge as such. The new engineer was rather a creature of public argumentation, his role being to propose new instruments, to fight tooth and nail over mathematical obscurities, and above all, to win enough money for the construction of new projects.[86] He was invented as a tool of the lobbyist, and his mathematics was designed to protect political lobbies.

The information in the minutes of evidence fell under two headings: one concerned with carriage design and another with road pavement. The approach favored by early committees consisted of intricate reports on carriage and wheel makeup that might have served the redefinition of the existing rules about carriages. The Committee on Broad Wheels argued that "any improvement in the construction of Wheels or Carriages would tend so much to the preservation of the Roads" that it was sure to "diminish, to a very considerable extent, the expences of their reparation."[87] Without raising taxes or altering the balance of local and parliamentary power, wheel engineering promised to drastically improve the state of the nation's roads.

A second approach consisted of suggestions for road surveying and building. This approach implied the need for a central government capable of making roads its concern.[88] The committee defined the problem as a mismatch between the "enormous weights" that "wagons are now allowed by law to carry" and the "materials of which our Roads are formed," which could nowhere, they argued, "bear the pressure." Either different standards for weights and carriages or different surfaces were needed. Sir John Sinclair, one of the lobbyists and a member of the committee, charged himself with recruiting prolific testimony from surveyors around the kingdom. He accumulated an alternate collection of "useful suggestions," also published with the report, that pertained not to carriage construction but to road pavement, stressing the superior standards of military construction and referring to the influence a general reorganization of parish authority might have if it regulated parish governments and their surveyors rather than targeting individual carriage drivers.[89]

At first, the balance between these two approaches was weighted strongly in favor of the broad wheels. Centralized spending other than for military reasons was unprecedented in Britain, and there was no civil authority with a bureaucracy large enough to take over the roads of the nation. The balance of argument would shift, however, with the special deployment of a corps of surveyors determined to wield their talents for the purposes of political change.

The process of committee work called out a new cadre of road builders distinguished in their aims and tools from the parish, turnpike, and military surveyors who had preceded them. Surveyors from parishes across Britain testified in Parliament about the benefits of local and national systems. They argued the benefits of systems that governed individual behavior in the form of wheel and carriage regulations versus state systems that ensured the pavement of roads across the entire nation. Surveyors and architects testifying exclusively about the benefits of nationalization gave themselves the name *civil engineers*. The term had been occasionally used to distinguish builders of civilian bridges and drainage projects from builders of military fortifications and roads since the seventeenth century and had been cultivated in the 1790s by John Smeaton to refer to his society of instrument makers and improvers. These improvers had a broad range of interests, including the construction of theaters, lighthouses, and steam engines.[90] But only when certain civil engineers began to concentrate on the immense government funding at stake for the roads did Smeaton's fledgling society start to develop the characteristics of a profession, with its own publications, tracts, licenses, and royal institute. Engineering took on a new role in advising Parliament about the potential costs and benefits of nationally constructed infrastructure.

MILITARY CRAFT AND PARLIAMENTARY EXPERTISE

These new publications were energized by the competition for parliamentary attention, remuneration, and funds. From the earliest professional textbooks, like Richard Lovell Edgeworth's *Essay on the Construction of Roads and Carriages* (1813) and Henry Parnell's *Treatise on Roads* (1833), these works defined the major calling of civil engineering in terms of the construction of the corridors of transport for the purposes of national trade and the road network in particular.[91] When the Royal Institute of Civil Engineers, launched in 1820, announced a new phase in the evolution of professional design, Telford, its first president, stressed in his speeches the political and national purpose of civil engineers. Telford was an exemplary sharpshooter of government funds, and under his direction, the Institute located itself in Westminster facing Parliament. According to Telford, "the vicinity of Parliament being almost essential to civil engineers for watching the progress of that peculiar but very important branch of legislation . . . afterwards carried into effect by them."[92] The new alliance between engineering and government made possible the consolidation of greater financial capital than ever before in advance of a road project's initiation. Civil engineers were distinguished from surveyors not by the field of construction, but rather by their talents of argument, presentation, and politics.

Richard Lovell Edgeworth, author of the first primer on road making, was an exemplar of the type: the political rather than the technical engineer. Edgeworth was a writer on education and a member of the Lunar Society, and his interests ranged from orreries to theater construction and carriage design. He began consolidating his thoughts on the relationship among government regulation, carriage-wheel acts, and road construction only after being solicited by Sir John Sinclair, member of the turnpike committee and chair of the Highland Roads project.[93] At this point, he turned his skills as a writer and educationist to the task of enumerating the various theories on whether wheels or roads were more important as a subject for regulation and delineating the varieties of military construction available to road builders.[94] Edgeworth's skills as a communicator proved more important than hands-on experience with road construction.

Similar virtues predominated among later civil engineers. John Loudon Macadam was a failed businessman who had attempted ventures in banking and tar manufacture in New York and Ayrshire before fleeing to the voluntary artillery corps and then to Bristol, where he allowed his gentlemanly duties as a magistrate and turnpike trustee to dominate his ailing business career. He helped the local lobby for a jail and served as an ordinary turnpike surveyor, and in 1810 he joined the caucus of other surveyors proposing their ideas before Parliament in response to the open invitation of Sinclair.[95] Macadam's didactic, spare style stripped military road building to its bare essentials in the form of an enumerated set of directions in the two-page "General Rules." It immediately served the needs of the parliamentary lobby for centralization, in desperate need of a simple how-to statement capable of codifying a single set of practices to be recommended to surveyors across the entire nation.[96] Macadam was overnight transformed from a failed businessman into the figurehead of expert engineering.

Thomas Telford, similarly inexperienced in construction, bolted from obscurity with the aid of political connections. Telford was an architect studying under the tutelage of Robert Adam when Sinclair enlisted him to design towns for the Highland Fisheries project, predecessor to the Highland Roads and Bridges. As Telford showed himself adept at drawing, explaining, and justifying the plans and accounts for government building, Sinclair entrusted him with greater projects: the plans for harbors and canals, the surveying of Scotland and Wales, and the creation of roads and bridges. When the projects received funding, Telford found himself in charge of a staff of trained surveyors, from whose ideas he learned the rudiments of paving and foundation laying according to the military style.[97] In the era of parliamentary road building, it was men skilled at explaining, enumerating expenses, and justifying work to be done, like Edgeworth, Macadam, and Telford, who rose to the rank of civil engineers. Their ability to work alongside parliamentary committees and to justify accounts and explain finances marked them out as the leaders in the new form of engineering.

The model offered by civil engineers, as opposed to the ordinary surveyors against whom they testified, was consistently one of top-down infrastructure at great expense, paid for by Parliament or a combination of parliamentary, local, and toll revenue and following the designs of a single engineer. These proposals differed from those of the surveyors in matters of permanence and expense: the civil engineering schemes imagined immense bridges and sixty-foot-wide highways that could be accomplished only with parliamentary money, whereas surveyors' proposals stressed local solutions that could be paid for out of local funds. The problem of permanence reflected the different models' relationships to money—federal versus local coffers—and also to the structures of assigning funds.

The traditional mode of funding infrastructure through local, piecemeal payment was ideal only for temporary or experimental forms of infrastructure. Local projects could be designed, tried out, redacted, and refunded at any given quarter sessions or meeting of the turnpike trustees. Parliamentary projects had to be funded by a single act of Parliament, approved by committee, recommended by the Treasury, and voted on in both houses. It was to the benefit of those seeking to build with parliamentary funding to front-load all their arguments in a single, overwhelming display of information, covering all possible expenses. Whatever was unconvincing or invisible in their recommendations might cause an entire project to be deferred to another meeting of the committee and defeated.

Techniques of Persuasion

Civil engineers developed two key techniques that enabled them to overwhelm the skeptics of parliamentary committees: the rigorous definition of general principles and the establishment of quantifiable metrics for projects to be executed.

From the first committees, civil engineers offered rigorous specifications of the correct formula for pavement in an attempt to establish themselves as authorities on road construction. The preliminary articulations were wide collections of the varied craft of road making as practiced across diverse parishes and turnpikes. "Observations Regarding the Formation of Roads" by Mr. Farey appeared in the 1806 minutes of evidence.[98] In the 1808 report, Sir Alexander Gordon of Culvenne in Scotland and Rev. Mr. Morphew of Norfolk pointed to the "great advantages which result from the making of Roads" according to the best methods available.[99] They pointed to the success of military road making both in Scotland and in recent areas where the troops had been deployed around Bagshot Heath. The centralizers explained the problem of certain "districts" where the making of roads had "not previously been much known"; in short, they proposed that government laws and "useful suggestions" might legislate the existence of basic roads in new districts previously served only by packhorses.

The most important of these general rules were those that applied to pavement. In 1811, the Committee on Turnpikes endorsed Macadam's suggestion of a single, broad standard of "well shaped and solid construction"— "a road made of small broken stone" washed from earth, ten inches deep. He recommended, in short, a gravel road, just like the ancient military ones beside which he grew up in Scotland, and his innovation consisted of suggesting it as a standard to which other roads could be held.[100] The suggestion of a single "general rule" for calibrating other roads essentially described a world in which the provinces all obeyed the vision of military coordination first set out by Defoe a hundred years before.

The rule answered Parliament's need for succinct statements of the diverse possible policies under review. It cut through the local variations in pavement practice demonstrated by expert testimony and offered, by way of consensus, a high ideal for pavement: the most permanent and indelible road that engineers could imagine.

The road proposed was also the most expensive. Within a few years of Macadam's celebrity, rival surveyors were pointing to the outrageous expense that macadamization involved when it was taken as the final rule, generalized without respect to local geography and materials. In the 1819 turnpike report, James Walker, the civil engineer employed by the East India Company on the Commercial Road and the other roads through the docklands, offered a different perspective. He explained that the quantity of materials necessary to form the road "depends so much upon the soil, and the nature of the materials themselves, that it is impossible to lay down any general rules for them." He described useful cases of chalk foundations (contrary to Macadam's preference for granite foundations on top of vegetable matter). By 1819, even Thomas Telford would suggest that Parliament save money by using local materials.[101] Macadamization, with its monolithic application of a single standard, made engineers blanch at a new fact: political expediency was establishing an ideal of construction far more expensive than it had to be.

There was also controversy over whether Macadam's distillation of military-style gravel roads represented the best possible technology available. Other engineers experimented with cement, wooden railways, and varying depths of foundations, each appropriate to regional variations in traffic and weather.[102] Telford recommended two general methods for road surfaces that differed from Macadam's. The first method was that of heavy flagstones atop a deep and stable foundation, generally applied to city streets. The second was a "deep foundation" of broken stones, gathered from the best local quarries, which Telford had proved could be both "more substantial" and "less expensive" than Macadam's model. When Telford's version was implemented on the Holyhead Road, the parliamentary highway became the only route in the area still passable during the "unusually severe frosts of the winter 1822–23."[103] Diversity of opinion rather than consensus marked engineers' own reactions to the principles of uniform general rules established by Parliament.

Indeed, Henry Parnell envisioned a cheaper, flexible system that emphasized local variations in materials and traffic flows. Centralized expert engineers working for Parliament, he believed, could identify the cheapest and most solid pavement appropriate to each locality. He reviewed the encyclopedic efforts of expert interviews as consolidated before parliamentary committees of the past, looking for local solutions. There was already enough knowledge accumulated by these efforts, he believed, to systematize the whole. For each kind of pavement, Parnell gave a history of its first introduction into use, the engineers who advocated it, and the reasons for consigning it to a certain geographic area and particular social use. He mapped the entire road system to systematize the roads according to traffic. "Roads of the first class" or "greatest thoroughfares" would be paved with enormous flagstones, while roads of the "second class" and the "third class" would ride on gravel foundations.[104] Together, expert tailoring of pavement to local materials and traffic flows could systematize the nation's roads for efficiency and cheapness.

Which general rule would prevail was not a matter that could be settled by science; it depended on the work of political debate. Telford's and Parnell's models offered economic efficiency. The virtue of Macadam's model, however, uniform standardization, ensured a relatively cheap mode of oversight. These were complex questions of expense, difficult to calculate. Was it more costly to measure each morsel of gravel in the kingdom from London, or to spread the most expensive gravel available across the whole of Britain's roads?

The matter was settled not by equations, but by lobbying. Macadam's rules were adopted by Parliament after the engineer conducted a massive publicity campaign on his own behalf. In a series of reprinted pamphlets in the 1810s, Macadam touted his travels "of twenty-six years on the roads of this kingdom" and explained his theory as the result of exhaustive knowledge of territory. His publicity effort was successful.[105] In 1825, Macadam was hired to pave Blackfriars Bridge with broken stone. In 1827, he oversaw the repaving of Regent Street and the Palace Yard with broken stone at a cost of some £12,000 for just over two thousand yards.[106] As was traditional, Macadam provided implicit refutation of the rival system, insisting that Telford's flagstone surface would "be too smooth for horses to go safely," having so little friction as to make them slip at an uncontrollable rate.[107] Publicity and publication ensured the success of a single system of engineering against its many rivals. Macadam's rules became the pattern legislated for all British roads in 1817.[108]

The victory of a monolithic system demonstrated the primacy of political efficiency over questions either of expense or of scientific knowledge. In 1817, parliamentary committees chose the expense of one ideal standard of gravel over the cheapness of scientifically calibrated, locally appropriate pavement. Systems of local materials, prescribed by local experts, would have reduced expense in the name of scientific learning. But Macadam's solution, the standardization of a general rule, offered an expedient solution by dispensing with the need for further calibration. By prescribing a single set of forms from London to which every surveyor in the realm must comply, it silenced the recommendations of science and hence the extension of debate. Where silence appeared cheaper than science, the "General Rule" became the favored tool of civil engineering.

A second set of tools used by civil engineers was the rigorous quantification of road standards to form evidence about the potential costs and benefits of road improvement. The first and easiest quality of the roads to quantify was distance. The surveys of current and proposed roads detailed the distance to be saved for every stretch of the journey and emphasized the new road as an efficient aid for later travelers. For the English section of the Holyhead Road, surveyed in 1819, Telford provided approximate costs of rebuilding hill by hill and inn by inn, explaining sections of road as short as three-quarters of a mile where the route avoided particular rises for every section of turnpike stretching between London and Shrewsbury. He included comparative tables illustrating for each of the fifteen trusts the miles for which it was responsible and its debt, interest, and tolls, and net return per mile.[109] The figures showed the revenue stream from tolls that parliamentary commissioners could expect to defray the expense of parliamentary building according to the most expensive and ambitious system of leveling and widening the road.

By listing each of the trusts along the road and computing their distances and costs as a sum, the distance tables also began the work of reimagining how the length of the road was managed across the boundaries of local jurisdictions. The sum total suggested that the road was already a single piece of highway rather than a route wending across a series of local jurisdictions. By showing the sum of the tolls collected from the public, the

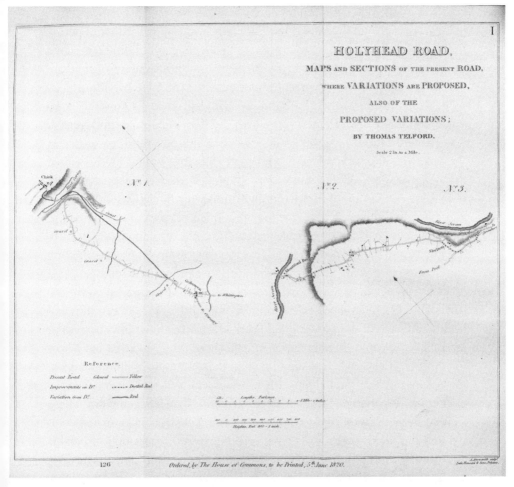

"Statement Comparing the Lengths and Revenue of Turnpike Trusts on the Holyhead Road," in Select Committee on the State of Roads between London and Holyhead, *Sixth Report*, 1819, p. 132.

distance tables imagined the tolls and maintenance as a single cost already weighing on the public. That cost to the public could be compared with any future propositions for centralized funding that would reconstruct the road as a single highway. Put to work in Parliament, distance tables proposed a quantified sum representing the present cost of the contemporary route and thus establishing a base against which future proposals for a consolidated highway could be measured.

Maps visibly demonstrated how the cost of building new roads would be offset by the potential benefit of obviating unnecessary distance. Their origins lay in the application of military trigonometry to the task of parliamentary argument. Measuring distance in advance of building entailed a new use of trigonometry: a transition from military sighting of roads to parliamentary plotting of current routes and their possible variations.

Sighting was performed by bands of soldiers moving across the landscape, staking out the clearest route with the help of boulders and trees. It was a means of finding a way—any way—through new territory.

In plotting, the surveyor used prior representations to find the best route through a district. Telford's many maps, presented to Parliament, were works of plotting, reexamining routes that the military investigations of the 1770s and 1780s had mapped. This method depended on paper, map, compass, and protractor to develop a reliable plan of the entire area, including its existing routes, and then to plot a more direct route across the landscape. Plotting could take place years before road building began, unlike sighting, which tended to proceed just ahead of the road being built. It depended on the abstracted and trigonometric calculation of exact distances between different passes, rather than the practical fieldwork of making direct lines across the land. Plotting could therefore approximate the shortest line that ran across the most even path, whereas sighting often took routes up and down treacherous slopes in an effort to follow the most direct line across territory.

Plotting also produced more material evidence than surveying did. These techniques were used to draw maps, sections, and charts that compared original and proposed routes, variations in the possible composition of heights and positions of drains, and different approaches to rivers and towns. The surveys had three objects, as Telford laid them out: to ascertain "the direction and shape of the vallies," to determine "the comparative elevations of the passes in the ridges which separate these vallies," and to determine the best crossings of the major rivers the route would pass, "particularly, in this case, the arm of the sea which divides Anglesea [sic] from Caernarson."[110] Unlike military surveys that charted the entirety of Scotland for the purpose of organizing later troop movements, the parliamentary surveys served as visual aids for argumentation in Parliament, illustrating expensive and unpopular plans for parliamentary road construction on particular roadways.

Official surveys of possible routes through Scotland were commissioned on behalf of the Committee on Roads by the Treasury in 1802. Telford's ambitious plan of Scotland in 1802 was the first extensive survey of that country since William Roy's laborious military survey in the 1750s. He explained that the military roads begun in 1742 had utterly decayed and were located "in such Directions, and so inconveniently steep, as to be nearly unfit for the Purposes of Civil Life."[111] As an alternative to repairing the military roads, he recommended plotting a new route running from Edinburgh to the north and east.

The potential of the plotted route as a persuasive technique ensured its patronage. Telford was commissioned to make surveys for other potential highways for the parliamentary committees on roads that followed. Between 1802 and 1830, Telford charted the territory for possible parliamentary projects connecting London with Dublin, Carlisle, York, Liverpool, Edinburgh, Roxburgh, and Glasgow. In each case, his maps clearly defined both the present route and the possible routes that would render "the Intercourse of the Country more perfect, by means of Bridges and Roads."[112]

In Parliament, the maps offered a visible and impressive demonstration of exacting labor put to the task of establishing accurate metrics for comparing costs and benefits. The presentation of the maps in committee ensured that they would receive a favorable reception as impartial and authoritative sources of information. Writers of the reports stressed the degree to which the maps were accepted as evidence of impartial reasoning. The Holyhead Road Committee boasted that Telford's engineers were "surveyors who had been accustomed to make surveys in hilly countries"— that is, engineers trained according to military surveying techniques as developed by the military surveyors of Scotland. They insisted on the accuracy of those methods, detailing the iterative "surveys and sections" and stringent comparison where surveyors' reports differed. The "Map, Plans, and Drawings" that synthesized this elaborate information were presented as an appendix to the report. Sections of 370 miles of road and surveys of 68 miles were offered, evidence, as Telford stressed, of "laborious perambulation performed." Judging "from the shape and distribution of the country" would be the only concern: the directness of the journey as opposed to the human interests of the communities along the route. The surveys that supplemented the maps further established the authority of their makers with an attentive and detailed account of each stretch of road along the way. The reports contained wordy descriptions of the soil, inclination, and terrain of the proposed route, sketching out features like "perpendicular rocks," "marly soil," and ascensions at an angle of "1 in 33." Each parcel of the survey contained estimated expenses of road making,

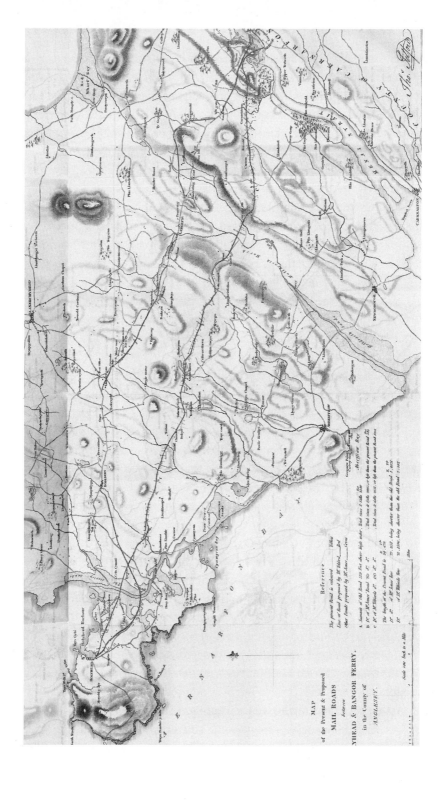

MILITARY CRAFT AND PARLIAMENTARY EXPERTISE

cutting, banking, parapets, bridges, cross drains, retaining walls, fences, land, and damages for the portion of the route in question.[113] Maps substantiated parliamentary claims by establishing the surveyor as a nonpartisan authority.

The maps ensured the victory of Telford's routes over those of other engineers. In 1809, he charted the best route between England and Ireland through northwest Scotland, identifying Carlisle as a potential port for improvement and target for new road making. The report included rival opinions of other engineers and local surveyors about the route and expenses estimated there, but the rivals included no maps. Telford's report included plans and elevations of no less than nine bridges across the Cree, Dee, Esk, Sark, and Eden Rivers. These drawings and descriptions stretched over sixty-three pages of the report.

To understand the persuasive power of Telford's maps, it is useful to consider the testimony of the rival surveyor interviewed in 1811. William Jones, a surveyor from Anglesey gave his testimony that although Telford's route was the most level and most direct, it would also be unnecessarily expensive, and another route practically as level and direct had been missed to the north. Jones pointed out that Telford sent the route across Lady Stanley's Sands, a waste of bad soils to the south of Anglesey. Jones preferred a more northern route above the Gwalchny Mountain, where "the materials [were] better," and the expensive importation of gravel could be avoided. But Jones's commonsense suggestions were ultimately disregarded because Jones had conducted no formalized survey on the basis of which to argue.

When John Maxwell Barry inquired, on behalf of the parliamentary committee, "How did you ascertain the ground to be as flat?" Jones answered, "By viewing the ground, and having been upon it many times." Indeed, Jones as a local could be expected to have a much more intimate memory of different locations than Telford.

Barry pressed. Had Jones drawn his conclusions from "any survey"?

"NO," responded Jones, "Not from any survey."

Here Barry jumped on him. "I supposed, that without a survey, you could not positively ascertain that to be the case?"

Thomas Telford, "Map of the Present Roads between Holyhead and Bangor Ferry," in Select Committee on the State of Roads between London and Holyhead, *Second Report*, 1819, p. 60.

Henry Welch, Surveyor, under the direction of Thomas Telford, "Map of the London and Edinburgh Mail Roads from East Retford to Morpeth" (London: Thos Telford, 1827). National Archives ZMAP 1/49.

Jones protested that indeed he had conducted many surveys, including most of the road under discussion, but he had prepared no map for this particular argument.

Barry had no use for Jones's method of reckoning. "If you thought that a more eligible line, why did you not lay it down in your Map?" the interviewer demanded.

Jones protested that he was following instructions to outline only the "present Road." A surveyor of the old method, Jones had sighted alternative routes and kept them in his head for the possibility of road making later. In the era of parliamentary debate, where paper evidence won arguments, traditional methods of practical road making without documents won nothing. It was visually displayed evidence represented in maps that per-

Figure demonstrating how to calculate the resistance a particular inclination posed to a vehicle, in Henry Parnell, A *Treatise on Roads* (1833), p. 440.

suaded politicians. Icons of long labor and detailed study in the field, maps helped establish the official surveyors commissioned by the parliamentary committees as legitimate sources of authority above local surveyors and other witnesses, persuading the rest of Parliament and the Treasury about the validity of the information they received about the road network.

A second point of comparison was that between the inclines of the steep roads before improvement and of the smooth roads proposed. "The most important part of the business of a skilful engineer," explained Parnell, was "to lay out the longitudinal inclinations of a road with the least quantity of cutting and embanking." Expert witnesses explained the significance of steep inclines in wearing out horses. In his *Treatise on Roads*, Parnell included tables showing how even slight differences in the rate of inclination made higher speeds of eight or ten miles per hour impossible for vehicles carrying any sort of load.

In the parliamentary committees, such theoretical examinations of the relationship between inclination and speed were made visible by personal accounts of specific territory. Official witnesses supplemented the evidence of maps with personal reports about the particular route over which most of these debates transpired. These accounts added a graphic layer of experience to warnings about the threats posed by steep inclines. The evidence for the Holyhead Road Committee included interviews with two of

	Height of summit above high water	Total rise and fall	Length	
			Miles	Yards
Old Road	339	3540	24	428
New Road	193	2257	21	1596
Difference	146	1283	2	592

Table showing the relationship among inclination, speed, and load, in Henry Parnell, A *Treatise on Roads* (1833), p. 48.

the Post Office's riding surveyors for North Wales, Christopher Saverland and George Western, who characterized particularly difficult stretches on the road, in particular, the steep hill at Betws-y-Coed, impossible to avoid, and pointed that a more "skilful surveyor would have avoided" the "ups and downs" of the current route.[114] These anecdotal illustrations of inclines could now be given exact numbers.

Quantifications of inclination proceeded first by numerical description and then by recommendations about a new standard. Telford took pains to note the difference in inclination offered by different routes. Parnell later explained the committee's suggestion of a national standard of "1 in 35" for "perfect safety" and a reasonable toll on "the labour of horses," and he held up the late parliamentary road building in Coventry as a model to be emulated. The ascents, Parnell stressed, were cumulative, such that in going from London to Barnet, 500 feet higher, a horse actually climbed 800 feet over the course of several hills. Parnell quoted Telford's reports, including tables, demonstrating how the leveling of hills resulted in a cumulative total of 1,283 feet saved over a short stretch of twenty-four miles on the road to Anglesey.[115] These numbers offered a confident description of the variety of road inclines throughout the kingdom and set up a preferred range of figures as an ideal.

To achieve the goal of reduced inclination, surveyors plotted possible routes with another set of priorities in mind, namely money. Smoother

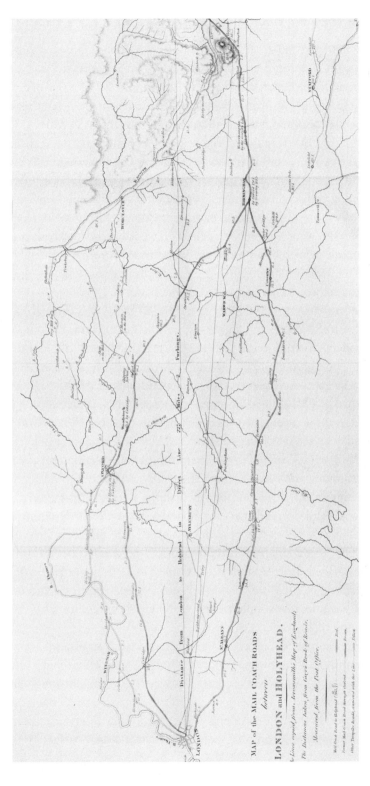

Excerpt, Thomas Telford, "Plan of the Holyhead Road," in Select Committee on the State of Roads between London and Holyhead, *Sixth Report*, 1819, appendix 6.

paths were better because they entailed fewer expensive cuts and fillings to level the path: as Parnell explained, a "gradual and continued" ascension required the smallest "quantity of earth to be removed."[116] The size of earthworks to be removed could also be quantified and compared, but this was a more sophisticated calculation than simply measuring distances overland. Comparing two routes that might connect Edinburgh to Yorkshire, Telford described his "very laborious surveys and calculations" and calculated the difference in rise and fall between the two lines.[117] He offered similar comparisons of two routes between Liverpool and London, in 1826, carefully remarking the necessary steep descent through Rugeley "at the rate of 1 in 25" that the less favorable path would entail.[118] These quantified comparisons were then rendered into a visual summary. Visual diagrams for each ten-mile section of the road highlighted the difference in distances of the present and proposed roads.

Even the smoothest route could often stand to be leveled through additional cuts of hills. These too could be quantified with enough labor. Surveyors provided a mathematical solution in the shape of new trigonometric formulas for quantifying and comparing the size of different masses before the cut was made. The exact angles and circumference of the earthwork were measured with the help of a new measuring tool, designed by John MacNeill, the resident engineer under Telford for the Holyhead Road between London and Shrewsbury, who offered his "Instrument for determining and setting out the line of slopes, in cuttings and embankments," in an independent publication that laid out the strategy. MacNeill published tables that simplified the geometric calculations necessary for estimating these figures and explained the "prismoidal formula" for calculating the geometric shape of the earthworks to be moved. The preface marked out particular tables as applying to the inclinations prescribed by Telford for the Holyhead Road, and the work was dedicated to Telford and the thirteen-year-old Institute of Civil Engineers.[119] Embankments, cuttings, and other earthworks became, like simple distances, a matter of trigonometry, easily quantified and compared.

Metrics of inclination lent punch to road advocates' arguments for centrally regulating the highways. Great inclinations were described as difficult for those going up, as well as dangerous for those going down. Parnell elaborated the dangers of "accidents by the overturning of a coach" and the "drags" put on the wheels of coaches trying to slow their speed, which inevitably resulted in "deep ruts" being made in the road surface.[120] The

John MacNeill, "Instrument for Determining and Setting Out the Line of Slopes, in Cuttings and Embankments," in his *Tables for Calculating the Cubic Quantity of Earth Work in the Cuttings and Embankments of Canals, Railways, and Turnpike Roads* (London: Roake and Varty, 1833), plate 4.

tables demonstrating the exact inclination of routes gave the centralizers and their engineers exact authority.

A third point of comparison was the nature of the road surface itself. Expert witnesses testified to the significance of friction in increasing the power required to haul a particular load. The 1836 turnpike committee interviewed Dionysius Lardner, author of A *Treatise on Geometry*, on friction. Parnell quoted the physics of Isaac Newton and James Woods to justify the relationship between the swiftness of vehicles, the hardness of the surface, and the inelasticity of the underlying foundation, comparing the carriage in motion to an iron ball projected over a "Turkey carpet" versus a "sheet of ice." His geometric diagrams illustrated how, he supposed, a carriage wheel sinking into an elastic pavement would have to overcome added resistance with every turn. The descriptive surveys were recounted as evidence about the deep ruts and rocky pavements that characterized many roads. Locals were asked to supplement these official records with personal accounts. Robert Latouche testified that he "found the Roads generally through Wales in a very bad and rough state, much worse than ever he experienced at the same time of the year, and so bad as to oblige the horses to go at a very moderate pace, lest the carriage should be broken."[121] General statements about resistance and the rockiness of roads provided the basis for civil engineers to propose a quantitative rubric for smoother surfaces.

The parliamentary committees even had quantitative information about the frictional resistance offered by different road surfaces. In the *Seventh Report* of the parliamentary commissioners of the Holyhead Road, MacNeill's "Road Indicator," a "machine for measuring the force of traction [sic]," established a reliable set of data about the power required for roads. It made calculations based on load and friction and allowed MacNeill to assert that the same load could be pulled by 156 horses in one place, but by only 56 horses in another; MacNeill, in other words, was measuring "horsepower" in terms of surface friction.[122] MacNeill explained that the instrument gave the commissioners the ability "to ascertain with accuracy and precision the state of any road, from time to time, as regards its surface." It pointed not only to the comparative condition of the road but also to "the amount of the deterioration, and the exact part of the road where such deterioration has taken place."[123] Having given figures to friction, engineers could promise to eliminate resistance by an exact science of pavement.

Debates over the best composition of foundations depended on advances in geology and the rigorous attention of surveyors to those findings. Charles Penfold, one of the surveyors interviewed, recommended basalts used as ballast in ships from China and Bombay, which had been "partially used in the macadamized streets of London." He also recommended the whinstones of Northumberland, the dark basalt of the Clee Hill in Shropshire, and granites from Aberdeen, Guernsey, and Dartmoor. Inadvisable stones included the soft granites of Cornwall and flints and limestones that predominated in the midsection of the country.[124] The entire science of geology was put to work in standardizing knowledge of the earth's contents across the map for the purposes of engineering. Surveyors benefited from geologists and became geologists themselves. "A fine tooth of the elephant" was found in the mud by one Mr. Stokes, surveyor of roads for Stratford-on-Avon, and it became part of the Royal Society's collection.[125] Pits of gravel, sand, and clay used for digging materials for roads served both the interests of paleontologists and the practical needs of surveyors.

During the 1820s, parliamentary and private engineers debated the best exact makeup of this underlying foundation. Mr. Wingrove, surveyor of the

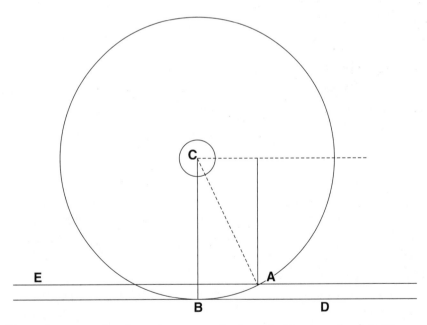

Figure demonstrating the resistance an elastic road offers to a wheel, in Henry Parnell, A Treatise on Roads (1833), p. 444.

Bath roads, used a combination of freestone brash and chalk for forming a foundation. Parnell advocated basalt, granite, quartz, syenite, and porphyry.[126] Macadam thought that the foundation should be "elastic" on vegetative matter so as to move the upper road surface with the movements of the earth.[127] Telford asserted that only a "non-elastic" foundation would prevent moisture from entering and disintegrating the stone within the foundation. MacNeill's machine for measuring horsepower confirmed that Telford's assertion had been true, and that the inelastic foundation produced less friction. Macadam's plan was dismissed as "very imperfect" and "very expensive," and Telford's inelastic foundation was set on the Holyhead Road and the Glasgow and Carlisle roads.[128] The measurement of friction and public debate together helped consolidate trust in the authority of the engineers.

Telford scientifically analyzed the makeup of foundations in an effort to rigorously and publicly assert the composition of an ideal model. Telford's first annual report on the Holyhead Road in 1823 gave tables for the thickness of the line of road between London and Shrewsbury, made by sinking holes into the road at short intervals. The thicknesses were found to range from $3\frac{3}{4}$ inches around Puddle Hill to $4\frac{1}{4}$ inches around Dunstable. The evidence was influential in convincing Parliament that a road of such importance should be equipped with a deeper foundation than the turnpikes were furnishing in their own right. Parnell asserted that turnpikes had an average foundation of three to six inches, and he recommended a minimum of six inches of gravel atop a deep foundation of larger stones.

The foundations were to be convex in shape, producing a convex pavement way with the purpose of draining water off to the side. Early engineers had experimented with various shapes, convex, concave, and flat, and had debated the worthiness of each arrangement in the pages of the parliamentary reports. Telford's books for engineering the Holyhead Road fixed the convex shape as the rule. Engineers were instructed in how to make the inclination of the tilt one to twenty. Levels were prescribed to aid each engineer in "giving a road a proper shape" and standardizing the curvature across the entire stretch of highway.[129] Through public debate and standardized figures, standards for road making were developed that fixed an ideal formula for pavement.

Studies of friction helped establish that a single form of foundation across the nation was desirable. MacNeill's report pointed to the "public advantages to be derived from such a system" and demonstrated in measurable terms the degree to which state intervention helped decrease expenditure for the civic body as a whole.[130] The details of its data gave a concrete, numerical, fungible portrait of the literal obstacles road surfaces posed to free-flowing trade.

Those surveyors who became civil engineers supplied a series of new technologies that differed in aim and kind from the road-building tools developed within the military. Their tactics for persuading potential supporters included an elaborate new series of devices for quantitatively measuring friction, mass, and distance. The figures were not meant to establish that parliamentary roads would be cheaper than turnpikes, but rather to fix exact numbers to the expense of road leveling and widening. In fact, these reports gave very little hard evidence to establish whether centralized or local highways were cheaper; instead, they lent an aura of efficient accounting to the centralizers' assertions.

Each of the technologies supplied numbers to the abstract notion of public expenditure on transport. With their maps of efficient routes and tables showing the minimization of drag, the accounts suggested that prodigious expenditures would be cheaper in the long run by reducing the horsepower of each individual user whose horse was saved time and energy by more even and smoother roads. They inserted into the ongoing clamor for cheap government another set of concerns having to do with the public and its relation to cheap trade.

Unlike the trigonometry and instruments devised by military engineers, the new inventions did not serve the purposes of building better; rather, they helped engineers argue better about accounts. The civil engineers' first forays were vague and clumsy. Proposing a bridge at Dunkeld over the river Tay in his 1802 survey of the Highlands roads, Telford merely guessed: "Under all these Circumstances the Expence would be considerable, and taking into Account the Uncertainty of the Foundations, the Amount cannot be stated at less than £15,000."[131] Gradually, engineers began estimating expenses more rigorously, applying trigonometry and new instruments for the purpose of accurately quantifying and comparing distances, inclinations, and road surfaces.

Although the figures became more precise, the reports nonetheless left glaring holes in their description of the proposed network as a whole. The parliamentary survey numbers actually told very little about the core issue, the relative costs and benefits of national versus local building. The engineers dealt with two very different kinds of numbers: first, quantitative distances and the projected costs associated with them; and second,

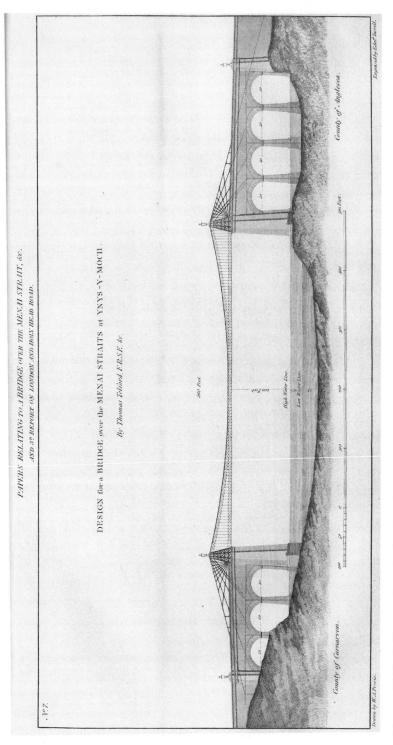

Thomas Telford, "Design for a Bridge over the Menai Straits at Ynys-Y-Moch," in Select Committee on the London and Holyhead Road, *Third Report*, 1819, appendix 7.

qualitative characterizations of the potential safety, comfort, and savings in horsepower each road offered. Despite the attempt to express comfort and horsepower as numbers, the qualitative and quantitative assessments could not be compared. No true cost-benefit analysis was possible with the tools at hand.

Nor were direct analyses of cheap government necessarily in the interest of the civil engineers. Their purpose was not to demonstrate financial efficiency as a fact, but to redirect the conversation from cheap government to cheap transport while establishing themselves as legitimate authorities who could be trusted with matters of national concern. Parliamentary surveying, unlike military surveying, fixed information about the relative costs of different procedures and invested those numbers with the authority of scientific inquiry.

Quantified figures, fixed to each leg of the territory in question, gave authority to the surveyors as potential executors of the project should it eventually secure parliamentary endorsement. Civil engineering's trigonometric formulas, diagrams, and tables offered symbols of authoritative control, much as the maps and surveys on the walls of gentlemen had done a century earlier. The equations and instruments associated with parliamentary surveying were political tools whose major use was to point to the skills of their users. They helped establish civil engineers as authority figures and so helped them gain the charge of the nation's highways.

The Characteristics of Standardized Building

Building with advanced planning was characterized by the production of a homogeneous swath of space around the highway that was made uniform across the entire nation without regard to local differences in geography or political structure. Civil engineers established themselves as authorities whose isolation from local negotiation was likely to help them better determine the most direct course of road. Trigonometrically plotting possible routes across maps, engineers tried to establish a possible standard above the concerns of a local community. Explaining his survey of the Welsh mountains for the Holyhead Road, Telford stressed how he "employed persons unknown in Wales" and "avoided communicating with Land Owners or others interested in the track through which the surveys were carried." Pointing to their national mandate and written records, civil engineers claimed to be distinguished from ordinary surveyors as impartial authorities worthy of parliamentary trust.

Once authority had been handed over to the civil engineers and their general rules, they set about standardizing the shape of the road itself. The Holyhead Road Committee required all footpaths to be about five feet wide, composed of "gravel and sandstone," and level with the center of the road.[132] Summarizing the rest of the committee regulations, Parnell advised that "foot pavements" should be of "well-dressed flags" two inches thick, tilting slightly toward the street so as to drain water from their surfaces, and "raised above the level of the pavement."[133] Indeed, Telford began adopting the strict analysis of road materials in his report, writing in 1819 for the Holyhead Committee, "The materials are gravel near Shrewsbury, and afterwards stone from Overly Hill; the latter is excellent, and should be used on most of the road; and if broken to a proper size, and the small stuff separated, and judiciously applied and made good, a very perfect road may be obtained." At the Wellington Trust, he put his assistant, Mr. Easton, in charge of working with the trusts, "selecting proper materials, and breaking and putting on stones." Around Wolverhampton he grew fiercely critical, complaining of "the irregularity of surface and direction" of the pavement, "being shaded by fences, and repaired with tender cinders, and ill broken stones." The Wednesbury and Birmingham Trust had "round pebble stones" that needed to be broken into smaller ones. Around Stratford he found "a coat of soft ill-cleaned gravel" with "heaps of stone lying unbroken, others broken imperfectly."[134] Telford was coming across the standard variations in material, use, and regularity that marked road building in most parts of England well into the nineteenth century. Civil engineers created a homogeneous landscape by holding different trusts to a general format for building.

The very landscape to either side of the road became part of a technological system standardized for use. Even the "hardest limestones" would "wear away very quickly when wet," so trees and fences would be pruned back to protect the road materials from accidents of weather and geography. Parnell quoted weather researchers on the "science of evaporation" to illustrate the persistent and pernicious effect of air moisture. He explained the reasoning for the wide, clear vistas of the new highways in his *Treatise on Roads:* "All woods, high banks, high walls, and old fences ought to be avoided, in order that the united action of the sun and wind may have full power to produce the most rapid evaporation of all moisture."[135] Parnell recommended a series of other standardized improvements, including the sodding of slopes, "retaining walls," supporting structures for roads on the face of hills, and raised foundations to lift roads that cut across flat or red land. He gave advice about compressing and draining the earthworks of retaining walls and embankments, comparing different techniques appropriate to clay, chalk, and sandstone. "In the Oxford clay, which covers so great a portion of the midland counties of England," he explained, "the slopes should not be less in any instance than two to one." Limestone, standing more solidly, would forgive slopes "at a quarter to one." He supplied copious geometric illustrations to advise the engineer about how to cut such embankments in different geologic strata. These general directions were supported by examples from the building of the Holyhead Road, in which Telford's advice was generalized for use throughout the nation.[136] The turnpike reform committees renewed forgotten legislation that required surveyors to erect milestones and guideposts "where two or more Crossways meet, with an inscription thereon in large Letters, containing the Name of the next Market-Town, to which each of the adjoining Highways leads, according to the Precept to him directed by the Justices at their Special Sessions for the Highways."[137] This seventeenth-century injunction had been left in the hands of the justices, who were sometimes reluctant to give instructions to thieves and highwaymen. In the era of parliamentary engineering, the legislation for milestones and guideposts took on new solidity. Standardized milestones and guideposts were provided along the whole of the Holyhead Road. Parnell summarized the committees' reports and their choice of "light-coloured stone, and of larger dimensions than usual," as well as their recommendation of the frequent use of "cast-iron posts" and cast-iron tablets mounted on stone. This legislation extended even to the dwellings provided for turnpike officials. Tollhouses should be "built in a strong and substantial manner, and made suitable and comfortable for the persons who are to inhabit them." The Anglesey tollhouse on the Holyhead Road offered a model for all that would follow. It consisted of a large octagonal lower room, paved with tiles for the floor, with bedrooms for the tollgate keeper and his family above. The whole was made of "good sound rubblework" with "hammer-dressed freestone, or slate" on the steps and sills, and Baltic fir for the inside timberwork. The inside would be "plastered three coats" and set in a pattern, painted white; the outside would be "roughcast, and coloured" a dark green.[138] The legislation of landscape, signage, and tollhouses perfected the homogenization of landscape into a single tract of space made uniform across the entire nation.

General rules and metrics of distance, inclination, and friction substantiated the link between expert oversight and public trade, displacing calls for cheap government and concerns about local power. As a result, a new regime of advanced planning was put in place whose major features were the burgeoning authority of a single expert in engineering and his amnesty from questions of local contingency, including later expenses that might arise over the long run, unlike military surveyors, who abandoned roads and bridges as soon as expense exceeded usefulness.

Despite the gradual change in the profession to correct for shortcomings in Macadam's formulation, the format of general rules ratified by Parliament continued to structure the relationship between engineers and localities. As a technology of trust, general rules were ideally designed to harness consensus and streamline top-down governance. They were easy to defend in committee and practical to enforce across multiple localities. Even when Telford's general rules finally replaced Macadam's as the standard for national building, it was still the format of general rules that Parliament recommended for the nation's highways.[139]

Civil engineering projects were frequently characterized by unforeseen expenses. Macadam's system incurred enormous expenses above the estimates Macadam originally offered, especially in the long term. Macadam's pavement in Regent Street and the Palace Yard incurred £5,000 a year for maintenance alone. Subsequent rates had proved that the savings in maintenance were less than projected. The House of Lords eventually did the math, and the newspapers reported that macadamizing, with its watering and cleaning, came out £2 per yard more expensive than Telford's more traditional method of flagstones. Macadamization had multiplied the cost of annual maintenance on Blackfriars Bridge by a factor of ten, with no compensating benefits. Gross unforeseen expenses could plague projects of such an immense scale, and once the decision had been made to follow a certain project, the authorities were burdened with the expense of either maintenance or reconfiguring the entire project.

The tendency was for these expenses to go unchecked until they became a scandal; otherwise, civil engineers kept reapplying to Parliament for further funding, creating a pattern of unsupervised expenditure. The costly maintenance associated with macadamized roads came to the attention of Parliament only after the expenditures of local governments burdened with the cost of maintenance were heavily advertised in the newspapers. The Holyhead Road began keeping track of these deficiencies only after 1820, when the English and Welsh roads both reported to

(No 3.)—Estimate of the Expense of sundry proposed Improvements between Bangor Ferry and Chirk Bridge, North Wales; February 1819.

Miles.	Yards.		£.	s.	d.
0	250	Hill at the Penrhyn Arms, Bangor, to be cut down	500	—	—
1	676	From the S. W. end of lot 24, past Ogwen Bank, to Ty-n-maes	1,399	16	—
1	110	From and including the bridge over the river Conway, near Glan Conway, to Hendre-issa toll-gate	1,951	11	—
3	768	From Hendre-issa toll-gate to Cernioge	5,221	1	—
0	1,000	From the S. E. end of the new road at Cerig-y-druidion	817	—	—
3	248	From the east end of the lot at Owen Glandwr's hill, to the west end of lot 19, near Rhysgog hill	4,512	3	9
0	660	Through the town of Llangollen	1,046	—	—
1	719	From Biddulph's limekilns, by Chirk Castle garden, to the top of the next hill	2,677	—	—
0	753	From Chirk, across the valley at the bridge	1,385	2	6
11	1,664	£.	19,509	14	3
		For expenses incurred since the operations were commenced, for surveys, acts of parliament, law expenses, superintendence and management, about	5,500	—	—
		£.	25,009	14	3

THE above is an Estimate of the Expense of the most necessary improvements still required between Bangor ferry and Chirk bridge; but as they cannot all be advantageously accomplished in twelve months, the sum of fifteen thousand pounds would be sufficient for the service of the present year - - say - - £.15,000.

"Estimate of the Expense of Sundry Proposed Improvements between Bangor Ferry and Chirk Bridge, North Wales, February 1819," in Select Committee on the Road from London to Holyhead (Turnpike Trusts), *Sixth Report*, 1819, appendix, p. 181.

the commissions. That year alone saw a deficit of £8,298 due to parliamentary fees, solicitors' bills, the salaries of the engineers, secretary, and inspectors, and other incidental expenses.[140] Telford's first constructions of the Menai Bridge suffered from numerous setbacks, and he reapplied for and successively was granted renewed sums of £52,000, £10,449, and £17,000 to keep the project moving. Once the original finances for a bridge across the Menai were supplied, Parliament proved reluctant to renounce the project despite new expenses that arose along the way. When the bridge collapsed in 1815, work was begun again. Lord Morpeth challenged the project in parliamentary debate, noting the gradual increase in expenses: "It appears that in the Sixth Report, in 1819, Mr. Telford estimated the road across Anglesey at £52,000; in 1832 there is an additional estimate to complete it of £10,449, and also a mixed account of £17,000 for the Menai bridge." Morpeth represented the financially conservative interests that balked at these large numbers, questioning the expertise that achieved parliamentary endorsement on the basis of one set of figures and then varied utterly from that projection.[141] But Morpeth's complaint came too late: the work of the civil engineers in projecting and estimating had secured them parliamentary endorsement in a streamlined process that dealt only with the original projection. After initial ratification of the project, later expenses were approved with amnesty. This lack of oversight and tendency to extend the project indefinitely, regardless of cost, differed radically from military building, where projects that got out of hand or were deemed irrelevant might simply be abandoned as ruins across the landscape.[142] In parliamentary projects, members of Parliament who depended on these projects as a list of their personal victories were unwilling to put a stop to the flow of funds to the engineers.

For similar reasons, engineers received amnesty from local political concerns. Because they were acting as representatives of Parliament on behalf of a supposedly national interest, the edicts of engineers were defended in law over the interests of local societies and even local governments. Eminent domain, which had first applied to allow local surveyors to make a roadway direct in the countryside, was now being applied to immense road-building projects in cities like London, where connecting the major highways required the tearing down of entire blocks of slums in St. Martin's-le-Grand to the north of St. Paul. Elderly men and women showed up to complain that they were losing their homes.[143] The case was unprecedented, for English law strictly protects the rights of householders and tenants. But the engineers' authority proved stronger; the complaints of the petitioners were thrown out, and the connecting streets were built, establishing the first major case for eminent domain in the modern world.

Civil engineers subjugated the landscape to the domain of a single rubric for practice. They streamlined the process of ratifying parliamentary endorsement of projects and made possible projects of greater scale and permanence than ever before. By investing builders with unprecedented funds and authority, parliamentary engineering consolidated command over local governments. It established new general standards that discarded information about local variance in favor of a single, enforceable standard. The result was a kind of civil engineering characterized by its deployment of large funds to alter landscapes without regard for local social or environmental context.

2

COLONIZING AT HOME

The Political Lobby for Centralizing Highways

> The border men, who used to make such furious raids and forays, have now . . . , like Telford, crossed the border with powers of road-making and bridge-building which have proved a source of increased civilization and well being to the population of the entire United Kingdom.
>
> —SAMUEL SMILES, *Lives of the Engineers*, 1861

In terms of the spectacle of modern engineering, the most impressive routes that a visitor to London in 1830 could take were the parliamentary routes to Scotland and Ireland. Setting out from the stagecoach yard before the gleaming marble temple front of the General Post Office in London, with the dome of St. Paul's looming overhead, the stagecoach would rattle over new-paved streets of smooth flagstones, only recently carved through parliamentary order from the former slums of St. Martin's-le-Grand, and then past the Angel Inn on Islington's High Street. Wide highways continued north, avoiding the imposing northern hills that had killed horses twenty years before and cutting through the new monumental canyon way of John Nash's soaring Highgate Archway. From there, the coach passed into the wide, smooth toll roads of St. Albans, bright with granite gravel. Within memory, these turnpikes, administered by local landlords formed into a private company, had been full of zigzags and potholes, but in the previous ten years they had been forced to conform to the regulations of experts appointed by Parliament, and now they were straight, smooth, and regular, with neat fences on either side and generous footpaths for pedestrians all the way. North of London, one could continue north toward Edinburgh or northwest toward Wales and the steamboats to Dublin.

These smooth pavements, soaring bridges, and steep embankments were built by parliamentary engineers, maintained by parliamentary surveyors, and financed through local divisions under the control of parliamentary commissioners. Parliamentary select committees, first appointed to look into the continued maintenance of Scotland's military roads, began to broaden their responsibilities in the first two decades of the nineteenth century, raising the question of centralized regulation of private turnpikes and local parishes.

Early visions of how centralized government could promote infrastructure preceded the birth of a lobby that envisioned the virtues of a new form of government, a unified bureaucracy overseen by experts. Its efforts culminated in the restructuring of local and turnpike government and the centralized building of interkingdom corridors between London, Edinburgh, and Dublin (via Holyhead) in the 1810s. The lobbyists pressed their suit on the basis of a variety of arguments having to do with national assimilation, financial efficiency, and military security.

The individuals who had the most to gain from these proposals were Celtic landlords situated far from England's centers of industry and exchange. Had they taken it on themselves to finance privately the highways that eventually connected the farthest Highlands and the western coast, they would have easily bankrupted themselves. Scottish and Irish landlords therefore turned to politics, arguing that the better integration of their markets would make Britain stronger as a whole. The lobby for centralized oversight and the highway system it built was almost entirely the work of Scots and Irishmen who retailored the structure of government in their favor.

This account explores the relationships between transport and the formation of the modern, bureaucratic state. Centrally organized highway building entailed a vast expansion of the state's centralized bureaucracy and powers over everyday life. The process of state road building in the 1810s and 1820s inaugurated an age of government collection of information, centralized standardization of local practice, the geographic redistribution of expenditure, and the creation of an expert bureaucracy, all factors that later became typical of nineteenth-century nation-states.

Traditional historians look to later events to explain the rise of Britain's bureaucracy. The era between 1790 and 1835 is presumed to be an era of cheap government, when concerns about the spiraling expenses of the military-fiscal state produced a political discourse about reform.[1] Most

British historians identify the rise of modern bureaucracy with the rise of social movements for sanitation and policing in the 1830s and 1840s, events that confirm the relationship between the rise of the modern state and the governing of the body.[2]

But by the first two decades of the nineteenth century, parliamentary road building was already pioneering new frontiers of scale, expense, and regulation. The advocates of government centralization introduced new tools for argumentation that involved reimagining data as a tool for political persuasion and control. The masses of data helped shift political discourse from concerns with expense to a new vision of how government expansion could spur political economy. Parliamentary road building thus succeeded in transforming the map of British trade, post, and defense, tying Scotland and Ireland closer to the centers of English activity. Even more important, these committees instituted a new form of bureaucracy that was subject to uniform regulations and characterized by the massive scale of vision and expense.

Although parliamentary roads never formed more than 2 percent of Britain's road network in length, the committees that controlled them diverted millions of pounds to engineering projects of unprecedented scale and ambition, driving corridors across gorges and into the underdeveloped fringe of the nation. In the process, they pioneered new forms of government centralization, expert rule, and technology. To prove the advantage of collective building over local practice, the committees demonstrated the benefit to the nation of turning over roads to a single engineer. The committees required their expert surveyors to produce mountains of charts, plans, and financial surveys that projected the roads' cost, thus introducing the technological skill set of the modern civil engineer. At the height of their success, visionary members of Parliament conjectured that a centralized board of roads could effectively administer the whole of the nation's roadways.

Four major kinds of roads therefore fall into periods punctuated by state investment in technology: between 1726 and 1750, a period of surveying, construction, and management innovation, driven by the military roads; between 1740 and 1760, a resulting turnpike and parish road boom; between 1795 and 1835, a period of parliamentary building during which there were innovations in route alignment, expert control of the roads, and the management of capital; and another period of parish and turnpike road building later in the nineteenth century. In all cases, state capital spurred the intensification of technology and its expansion to Britain's geographic edge, while private and local administration disseminated that technology to the territories of inland England.

By the nineteenth century, any traveler could identify two major types of roads in Britain, each characterized by a different relationship to government. Military and parliamentary roads, directly financed by the government, were the major engines of technological innovation and stretched across a newly connected, national space. Local and turnpike roads, financed from local taxes and toll collection, aimed to satisfy local needs.

The advantages of each system rested on different relationships between region and funding. Celtic landlords competed with English ratepayers for development funds on which their access to markets depended. To many, the emergence of regional lobbies and state-endowed infrastructure grants was a harbinger of serious problems.

Defoe's Vision

By the end of the seventeenth century, Britain already had a formula for how a centralized system of road administration might work. Daniel Defoe's visionary treatise, "Of the High-Ways," formed part of his *Essays upon Several Projects*, published in 1696–1697 in the fit of idealism and experimental thinking about government in the aftermath of William II's rise to power and the end of war with France, a period Defoe referred to as "the Projecting Age."[3] These were the far-reaching visions of a nonconformist and partook of a utopianism and experimentation allowed for by the window of opportunity of a new administration, a window that would be quickly shut in the first decades of the eighteenth century with the return to other forms of business.

Defoe was writing in the midst of shifting ideas about travelers, who in the 1690s were increasing not only in numbers but also in prestige. State-deployed excisemen were constantly on the road, and MPs from Scotland and Ireland traveled more than their predecessors before union. The expansion of the Post Office into a regular system of delivery had similarly dramatically increased the volume of official travelers on the road, and the numbers of private stagecoaches were rising in the 1730s and 1740s. All these forces tended to suggest the existence of a larger community of travelers who needed to be protected from the greed of bridge keepers and tollgate collectors. The responsibility for mediating these relationships fell on the community and the state. Between 1650 and 1750, both courts and Parliament began to protect the traveler from gaping holes in the middle of the road, ruling that local surveyors around the kingdom had a duty to fence them and then fill them in.[4] The rights of travelers were becoming established. But how far collective efforts to manage the roads were reasonable, within England's system of local rule and well-protected liberties, was still unclear.

Defoe devised what seemed like a straightforward solution of a national administration of the roads that would supplant local inefficiencies with a single project of technical management. He proposed "that an Act of Parliament be made, with Liberty for the Undertakers to Dig and Trench, to cut down Hedges and Trees, or whatever is needful for ditching, dreining and carrying off Water, cleaning, enlarging, and leveling the Roads."[5] Large-scale coordination would entail new measures of efficiency that promised, in the end, to be cheaper for individual rent payers. Defoe explicitly avoided the suggestion that state administration would mean new roads; rather, it should improve and make more solid, and thereby simply cheapen, existing roads. He envisioned "a noble Magnificent Causeway" "Four Foot High at least, and from Thirty to Forty Foot Broad, to reach from *London* to *Barnet*, Pav'd in the middle, to keep it Cop'd, and so suppli'd with Gravel and other proper Materials, as shou'd secure it from Decay with small Repairing," a "Fabrick" that would need less frequent repair. Crucial to Defoe's vision was the skilled deployment of the best technological methods available to his contemporaries. He described in detail the forms of ditches and culverts that could best divert water from the road surface. If collaboration were effected on so grand a scale, a variety of innovative and efficient improvements could be coordinated. Borrowing from the improvements implemented by Peter the Great at crossroads across Russia, Defoe envisaged communally owned "cottages at proper distances" such that "a Man might Travel over all England as through a Street, where he cou'd never want, either Rescue from Thieves, or Directions for his way." Defoe pointed to the extended improvements that could be effected in travel, given a cooperative community with certain powers, labor, and money. The possibilities of improving English trade indeed seemed remarkable. Imagining these improvements, however, required political and financial reckonings of some consideration.

Defoe envisioned minimizing costs by distributing the work of road maintenance at the parish level. The routine tasks of road upkeep would be distributed out of a national fund by officials appointed in every parish.

Each of the communal cottages serving as inns would be tended by a poor man of the parish. To this local system, however, Defoe proposed adding a mixed system of centralized oversight, managed by the state. The general condition of the roads would be watched in every county by "Two Riders" who operated on circuits in much the same way as excise collectors, "always moving the Rounds, to view every thing out of Repair, and make Report to the Directors, and to see that the Cottagers do their Duty." Two other maintenance men in every parish would cleanse the ditches and fill in potholes. Although road building would have to be centrally coordinated by a new government body, road maintenance would be restored to the hands of the parishes, where it would prove to be a more enlightened version of the contemporary system.

Defoe had in sight two types of bureaucracy that could be trusted with advanced improvement projects on a large scale: the French corvée of forced labor and the Bank of England. His vision combined both seamlessly. Fifteen "Undertakers" from every county and ten trustees would form a commission with the power "to press Wagons, Carts, and Horses, Oxen, and Men, and detain them to work a certain Limited Time." Additional labor for first erecting a permanent road system would be supplied by convicts, especially the prisoners of the Old Bailey who were sentenced to death or transportation overseas for lesser crimes. The forced labor of convicts would be supplemented with that of "200 Negroes, who are generally Persons that do a great deal of Work," leased from the Guinea Company. Even the rent on a slave corvée, however, was more than early modern states were equipped to provide. Defoe therefore suggested that the expense of overseeing slaves be funded from a national tax on land administered by an independent bank in Middlesex. If this sum proved insufficient, Defoe reckoned, it could be supplemented by the enclosure of the wastelands that bordered the rights of way, the sale of which would generate a "Prodigious Stock of Money" to be managed by the Bank of England and the "Publick Stock" for the purpose of continuously managing the roads. Defoe's suggestions, in their complexity and their tyranny, suggest the degree to which road building required a scale of finance never before contemplated.

Defoe intentionally minimized the issue of labor to make the issue seem simpler than it was. Nominating an army of two hundred "Guinea slaves" and convict laborers allowed Defoe to skirt the massive problem of supplying, funding, and managing so large a civil corps in the absence of coercion. It was hard to see exactly how the management of such labor would work, and the potential objections were manifold. English travelers in Sweden noticed how efficient the centralized system of roads and post horses was, but upon inquiry, they objected to the Hollskjuts, the national system of statute labor, as "extremely burdensome to the natives."[6] English writers negatively compared the French corvée, "an intolerable burden," with the English duty of statute labor on the roads, limited "under proper restrictions."[7] Arthur Young, traveling in prerevolutionary France, was struck not only by how the corvée kept the peasants "poor and depressed," but also by the variety of complaints against the corvée in Breton folk speech, "unknown in England, and consequently untranslatable."[8] Calls for reform of the English system of parish statute labor tended to encourage not a more intense system of labor, but rather its replacement by whatever form of tax assessment would best minimize the impact of road maintenance on the population.[9] When they considered the labor dedicated to the maintenance of highways, English writers were proud of the effective limits on their government.

In the English system in 1689, no authority comparable to the one Defoe imagined existed that was capable of commandeering labor or overturning property rights. The system was therefore limited in how far it could project, how expertly it could build, and how rationally it could plan the map of expanding highways. It did, however, fiercely protect those individual liberties enshrined in the English Bill of Rights. The evolving system that existed, in the absence of Defoe's board of roads, was rather a patchwork of public and private road management on a local level, substantially constrained within the rule of law.

Defoe's proposal imagined measures for forced purchase and realignment of land utterly without precedent in common law.[10] Defoe conceded that popular opinion was not in favor of such an expansion of state power, "because the liberty seems very large, and some may think 'tis too great a Power to be granted to any Body of Men over their Neighbours."[11] His suggestion went against the precedent and practice of British law.

In the system of local parish and turnpike regulation, the powers of diverting roads were strictly limited. The Tudor statutes of 1555 instituted a slate of basic requirements that parishes were supposed to enforce for their own highways, obligating authorities to keep the road clear of impediments; these statutes were applied to turnpike roads as well. A public jury was supposed "to assess the Value of such Ground, not exceeding Forty

Years Purchase of the annual Value thereof, together with Recompence for making Fences and Ditches by the Side thereof," and it was empowered to settle with the owner in advance of building.[12] Within this local system of activity, Parliament's role was limited to rubber-stamping roads and bridges approved within the sphere of local consensus.

Despite these constraints, the local system was still susceptible to corruption, and greed sometimes prevailed over turnpike trustees despite the watchful guardianship of local landlords. In 1781, Mr. Bargus and Mr. Mant of the Fareham to Gosport Turnpike seized part of the bishop of Winchester's mudflats to widen a narrow road passing between a sluice and a bridge. They were taken to court by the bishop and a quay owner whose livelihood had been ruined thereby and were made the subject of a chastising pamphlet that sarcastically mocked *"their zeal for the public good!!!"*[13] The seizing of the bishop's property without notice was clearly illegal, and there were good grounds for supposing that cutting off the quay was unjust as well. Strictures limiting the profit of trustees and requiring juries to assess property before a turnpike trust could build new roads assured that the ownership of private property would be kept sacred, despite the temptations of land speculation necessarily posed by road development. The turnpikes and local roads were thus constrained by respect for property rights of a kind that defied Defoe's vision of a powerful and centrally managed highway system.

The limits on turnpike building sharply restricted the activity of local building. Turnpike trusts, protected by law, regularly fought other trusts whose projected roads would diminish their own revenues. In 1761, the Totnes Turnpike trustees in Devon brought a suit against the Newton-Bushell Turnpike trustees, who had begun building an alternative branch that would compete with the recently completed Totnes road. The Totnes trustees complained that there were only enough travelers to support one of the two trusts and begged Parliament to suppress the newcomers.[14] By granting particular turnpikes and not others, Parliament defended the monopolies of earlier investors. Turnpikes were also subject to tedious legal arbitrations intended to ensure the protection of individual property rights. Indeed, contemporaries joked that settling such fights was the only thing justices of the peace had time for. In 1786, a "Mrs. Bustle" complained that her husband, "Mr. Bustle," recently a justice of the peace, "now does little else but study law-cases, convene meetings about highways, turnpikes, bridges, and game-licenses, and ride all over the country, dispensing justice, redressing wrongs, removing nuisances, and punishing delinquents." Nor was this exertion toward settling arguments met with public accolades. She complained, "There have been such bickerings amongst the gentlemen about the widening of roads, removing of dunghills, pulling down cottages, and punishing of vagrants, that one half of the neighbours are scarce in speaking-terms with the other."[15] The turnpike and parish system protected individual liberties at the cost of frequent meetings and limited opportunity to plan the transport system as a whole.

The turnpike and parish system was also limited by its local and regional focus. These roads reflected the development only of markets already sufficiently powerful to fund roads. Unlike the powerful state imagined by Defoe, they were incapable of extending commerce to new districts. Turnpike trusts paved the major spokes of London in the 1720s, a few major English arteries in the 1750s and 1760s, and the byroads around London, Manchester, Leeds, and Newcastle between 1790 and 1820.[16] But these regional developments around England's most powerful markets left most of Wales, Scotland, Ireland, and the nation's coastal fringes connected by ancient trackways that were unpredictable at best.[17] Once the traveler left the zones of commerce, the underdeveloped routes to the periphery were closer to animal trails than roads.

The turnpike and parish road system was gradually amended during the eighteenth century by various relatively ineffectual acts intended to enforce a minimal standard of road management across regions of vastly different levels of development. The laws of the 1750s and 1760s set forth requirements on axle length, the number of horses, and the removal of trees from the road.[18] In general, these acts made individuals, rather than parishes, beholden to the law. Individuals could be indicted before a local justice, while the indictment of parishes and turnpikes would have required accountability to a centralized authority that simply did not exist. As a result, individual travelers were managed within a wide net of surveillance that pretended to hold them responsible for not destroying the road surface for others. The local system, with its regulations for individuals, aimed to eradicate the worst ruts and potholes, but there its power ended. Local legislation could scarcely conceive of a plan designed to bring substandard routes up to par; it offered nothing for the poor parish and nowhere aimed to develop a region that lacked access to markets.

Defoe's vision posed a challenge for Britons who were willing to speculate about a more direct, better-managed, and better planned road system under centralized direction. For much of the eighteenth century, it remained an unparalleled and unapplied conjecture about what capital could do for development. Only in the 1760s and 1770s, as Britain's empire in the New World fell apart, did Britons begin to look for an economy that rested on development at home. Writers looking for domestic routes to prosperity would return to Defoe's vision and describe the variety of reforms that might free the circulation of commerce at home.

Systematic Problems

The new forms of government that appeared in the era of Defoe encouraged many to contemplate a more just allocation of government. If Defoe envisioned the state provision of hostels for travelers, others imagined a government that cared for the poor at home rather than waging expensive wars across the Atlantic. But in the opening decades of the eighteenth century, the forces of traditional rule grew stronger. At the national level, money and policy were increasingly oriented toward expensive wars on the Continent and in North America. At the local level, fewer Britons were allowed to participate in the parish meetings where poor-law rates were decided. More decisions were being made in distant courts.[19]

Political imagination of an alternative order, one that recognized the interests of the many, gradually coalesced into a strain of radical politics capable of challenging the direction of government as a whole. In the 1760s, a rising tide of "politics out of doors" began to confront the military-fiscal state's increasing and expensive adventures.[20] The efforts of a new generation of ethicists and political economists had expanded the concept of the "public good" to emphasize the interstices where the wealth of nations intersected with care for the poor. Theories of the 1790s about solving the vagrancy and unemployment problems together emphasized the role of statute labor on the roads as an example of how the shared economy could work on behalf of all.

The roads were the one setting where utopian ideas were regularly translated into reality. From early in the eighteenth century, vicars and squires had regularly executed new ideas about improvements at the local level. Through their work, England was already enjoying tangible benefits in speed and regularity of travel by the 1760s. At the same time, the influence of continental philosophy was pressing squires to expand their vision. Thinkers like Adam Smith and Bernard Mandeville urged their readers to consider how the same improvements could expand if they were executed on a grander scale. Influenced by political economy, a new generation of writers began to spell out the benefits that would accrue from a systematic network of improvements.

Local vicars and gentlemen had worked in the business of small-time improvements for a long time. Sharing treatises on new methods of raising turnips or planting hedgerows, they tended to think in terms of long-term improvements easily executed by a single individual or a small group of gentlemen working together in a parish. Many of the gentlemanly administrators who staffed the boards of turnpike trusts considered themselves attuned to the public interest and capable of pointing out faults in administration whose correction would produce a better road system overall. Individual trustees such as Henry Sacheverell Homer, a Warwickshire vicar and turnpike trustee, examined the shortcomings of the turnpike system from the vantage of local commerce and regulation.[21] Their treatises typically advocated piecemeal and voluntarist reforms. They stressed the inefficiencies related to interest and monopoly, features that could hypothetically be restrained by the provision of a centralized body with the power of oversight.

It was at this point that economists threw down a challenge to gentlemanly readers. Whatever improvements could be accomplished by an individual could almost certainly be generalized in scale across a nation if the laws of economic growth were discovered and exploited. In France, writers like Turgot and Condorcet advocated the search for the abstract laws that governed economic growth. The physiocrats defined their quest as the marriage of physics (actual observation of the rules of nature) with sentiment (sensitivity to customs and ethics in social life). This search led them to emphasize the importance of state policy in enabling economic growth. Tariffs, complex tax laws, and currency manipulation imposed a foolish, artificial order on the economy. If low taxes, minimal police repression, and strong communications held a nation together, they urged, the natural outgrowth would be the increase of prosperity for all.[22] For English economists influenced by the physiocrats, the free market's behavior clearly required a strong communications network. The failure of communications, warned Adam Smith, was expressed as "monopoly." Monopoly's high prices kept potential actors from participating in the market and thus operated as "a great enemy to good management." Only good roads could ensure that diverse classes of producers had an equal opportunity to participate. If Parliament were to build long roads, it would "introduce some rival commodities into the old market," with the result of "breaking down the monopoly of the country."[23]

By opening up access to trade, roads would inaugurate a truly free market, characterized by wealth and opportunity for all. Throughout the eighteenth century, critics of the turnpike and parish system noted the inefficiencies of Britain's roads, where "a poor Traveller . . . at every Ten Miles End is stop'd by a Turnpike, and dunn'd for a Penny for mending the Roads in the Summer, with what everybody knows will be Dirt before the Winter that succeeds it is expired."[24] Cheap roads were a necessary foundation of free markets.

In more recent times, we have come to see eighteenth-century economists like Mandeville and Smith as critical of the state. In fact, these authors laid the foundation for considering how states could nurture their domestic economies by ensuring infrastructure. They drew attention to the shortcomings of local management in the turnpike system. Mandeville, who elsewhere so strongly encouraged individual initiative, argued in the case of highways that a collective approach would be better. Only autocratic states had turned their basic infrastructure into "works of Duration."[25]

Britain, Mandeville maintained, could benefit by following the example of great empires and strong states. "Let us look back on the Stupendious Works of the *Romans*, more especially their Highways and Aqueducts," he urged. "Let us consider in one view the vast Extent of several of their Roads, how substantial they made them, and what Duration they have been of."[26] Smith lamented how Britain's roads lagged behind those of other advanced nations. He pointed Britain toward the economic success enjoyed by despotic governments capable of extending good highways across their territory: "In China, and several other governments of Asia, the executive power charges itself both with the reparation of the highroads, and with the maintenance of the navigable canals. . . . In France the funds destined for the reparation of the high roads are under the immediate direction of the executive power."[27] Such essays remained a testament to the lasting power of Defoe's vision, demonstrating that even those who served on turnpike trusts recognized the inherent limits of the system and glimpsed the ways in which centralization could alter the landscape in a way impossible for local administration.

A more radical vision was offered by those who had the most to gain from a system as aggressive as that described by Defoe. In the 1790s, Scottish landlords began to conceive of a centrally funded and administered road network that would bring great highways to the undeveloped Scottish periphery. John Sinclair, an Ulster landlord and correspondent of Adam Smith's, began envisioning a new role for centralized government during his tours of the Continent in the 1780s and pioneered the modern collection of data for political argument.[28] Sinclair conducted a prolonged correspondence with surveyors and turnpike administrators around the nation and published his findings in his *Statistical Account* between 1793 and 1798, in which he identified the failures of the turnpike system.[29] After 1794, Sinclair arranged for the newly formed Board of Agriculture to begin collecting data on a set of ideal standards for road management based on the experience of individual surveyors.[30]

Since his adolescence, Sinclair had pursued an interest in Scottish nationalism through poetry, translation, and song, developing a passion for the poems of Ossian and the preservation of Gaelic. In his twenties and thirties, having inherited his father's estates, which covered one-sixth of the county of Caithness, Sinclair sought a new means of advocating his nationalism, based not on literature but on the economy. Travels in northern Europe in 1786 impressed him with the strength of centralized administration in Prussia and filled him with "ardour to establish, in my own country, all the beneficial institutions which were scattered over others."[31]

Sinclair's fact-finding missions allowed him to publish the viewpoints of individual trustees and surveyors discontented with the system of local management. John Wright of Chelsea, a turnpike trustee, argued that a national body was necessary to ensure that the trusts' funds actually went to improving the surfaces of roads and proposed an "office in London . . . to receive and audit the accounts of trustees."[32] Sinclair had also written to John Holt, a Liverpool schoolmaster who had acted as a highway surveyor, and Holt was eventually commissioned to perform the agricultural survey of Lancashire.[33] Holt urged "the revision of the general law" as fundamentally "necessary."[34] Among the Board of Agriculture's correspondents was Robert Beatson of Fife, an army officer and agricultural improver who had also corresponded with Adam Smith.[35] Beatson proposed "a board of roads and internal communications" that would be invested with "the controlling power of management of all the public roads and canals in the kingdom; the letting of tolls, or collecting the revenues arising from those roads and canals; . . . and, in short, the whole power of regulating and deciding every thing respecting so important a trust."[36] These men urged that centralized government, armed with expert guidance, could link the Celtic periphery to the rest of the nation, replacing expensive exploits abroad with the cheaper alternative that Southey called "colonization at home."[37]

An even broader vision of a truly national road system was championed in the years that followed. During the 1810s, Parliament designed and financed an entire interkingdom highway system, connecting London to its colonial capitals of Dublin and Edinburgh. These parliamentary committees, like Sinclair's, were designed by another Celtic landlord, the Irish MP Sir Henry Parnell. Parnell directed the series of parliamentary committees that appointed new boards of highways ultimately responsible for repaving the Holyhead Road and parts of the Great North Road in Wales and England.[38] By the 1830s, plans were mooted for bringing all of Britain's roads under the guidance of a single authority. More than a hundred years after the publication of Defoe's treatise, Parliament finally inaugurated in the select committees an institution capable of carrying forward the vision of centralized infrastructure.

The highway lobby was supported by a broadening army of Scots and Irish MPs. National spending on the improvement of former colonies was potentially good not just for the Scots, but for the Irish as well, who had represented their nation in Parliament since union in 1800. Since the first activities of the Highland Society, Irish landlords had been impressed with the Scots' success in prevailing on English charity. Colonel Tittler weighed the question of Irish union in a pamphlet titled "Whether Scotland Has Gained, or Lost, by an Union with England" and argued that highways were a major potential benefit of union.[39] Irish landlords reasoned that if Scotland had benefited from English welfare to such a great degree, Ireland could secure highways as well. Scottish activism began to serve as a model for Irish landlords, who readily patterned themselves into a new lobby for centrally built roads. The Highland Society began corresponding with sister institutions in Ireland that aimed at encouraging centralized promotion of roads to Ireland, including the Dublin Society and the Farming Society.[40] Irish landlords in Parliament like Henry Parnell, MP for Queen's County, began directing select committees for a road that would connect Dublin with London.[41] The Holyhead Road Committee, like the Highland committees before it, was stocked with those MPs most interested in volunteering for the hours of presentation of evidence required of committee members; it was consequently in the hands of Irish, Welsh, and Scottish landlords.

Together, Irish and Scottish MPs could stack the parliamentary committees for roads in their favor more effectively than the Scots could on their own. On the first Highland Roads and Bridges Committee, responsible for securing initial funding over the course of the earliest four reports, the Scottish interest was represented by William Dundas and Mr. Grant, both of Invernesshire.[42] The 1809 committee that reported on Telford's proposed route to connect England and Ireland at Carlisle was heavily weighted on the side of English financial reformers: Nicholas Vansittart of the Treasury; Hawkins Browne; Sir Robert Buxton; and William Smith of Norwich, a banker in London. On the Celtic side were Sir James Graham and Sir Willam Pulteney of Dumfriesshire, MP for Shrewsbury, a correspondent of John Sinclair's and Adam Smith's.[43] By the time Irish MPs entered the highway game, in the 1810s and 1820s, their presence weighted the committees in favor of the Celtic interest. In the 1820s, Sir Henry Parnell was joined on the Holyhead Road Committee by four Welsh, Irish, and Scottish representatives, including Scottish members standing in for the Treasury.[44] In the controversial 1824 Holyhead Road commission, which voted to centralize the local, English turnpikes, six Celts were balanced against six Englishmen, one of whom, Davies Gilbert, the Oxford mathematician, was a constant proponent of advanced engineering.[45] Alliance building between Scots and Irish weighted the committees in favor of highway reform.

The chief instruments of political conquest were publications. The administrators and their engineers—Thomas Telford and John Loudon Macadam—published innumerable parliamentary reports outlining the history of development under turnpike roads and the case for centralized development. These preliminary reports were then expanded into widely circulating treatises on road making and the importance of centralized government. Macadam's *Remarks on the Present System of Road Making* was published in 1823, Parnell's *Treatise on Roads* in 1833, and Telford's autobiographical manifesto in 1838.[46] These texts repeated the arguments that centralization would eradicate the financial inefficiencies of the turnpikes, facilitate national integration, and contribute to traveler safety and national security.

In the debates that followed, the centralizers found an ally in the Post Office, which since 1784 had been indicting parishes wherever sticky roads delayed mail coaches on their strict timetables. The Post Office had no authority to force parishes and turnpike trusts to perfect their roads, but its surveyors regularly testified before Parliament about the needfulness of new projects that promised to improve the condition of the roads overall.[47] Post Office surveyors collected painfully exact timetables of stagecoach routes, the first quantified testimony about the national variability of road conditions.

In the hands of the parliamentary committees, these timetables and surveyors' testimonials became key to arguing the necessity of improved roads across the entire nation. By 1808, Sinclair persuaded Parliament to appoint a standing committee, chaired by himself, to look into the state of roads in the Highlands. With the help of Post Office data, Sinclair's committee inaugurated the building of hundreds of miles of new road, the first expert-built and centrally administered network of modern roads intended for civilian use.[48]

The committees interviewed local administrators of turnpikes, Scottish landowners, and parliamentary board members about potential improvements to the nation's roads. Turnpike administrators generally confined themselves to speculating about the problem of local monopolies and conjecturing that a centralized authority could enforce the best practices for financial efficiency and accountability. The Scots added arguments in favor of expert engineering, coordinated from London, and stressed that well-built roads could extend prosperity to the nation's fringes. Finally, members of Parliament allied with Scots and Irish landlords urged other MPs to join their cause, pointing to roads as a tool for cultural assimilation, military reconnaissance, and national security. Together, the arguments for financial efficiency, expert oversight, economic development, and national security persuaded Parliament to pass the radical vision of a centralized road system.

Financial Efficiency

At the close of the Napoleonic Wars, Britain was a nation desperate to cut its expenses. The massive public debt and fierce taxation of 1815 had generated reactions from all sides. Luddite riots, Henry Hunt's orations, and radical petitions all added weight to Henry Brougham's colonization of the term "reform" on behalf of Whiggery, countered only by stoking the aging but sharp Pittite rhetoric of "reform" among the Tories.[49]

The lobbyists who assembled to demand highways were asking for a massive increase in parliamentary expenses at a time when nearly everyone was working to cut government spending. Their project depended on persuading Parliament and members of the public that better roads really would lead to greater riches for all.

Systematic control cut costs in the long term: this was the argument on which road advocacy depended. In the 1790s, writers for the Board of Agriculture urged that pooling expenses among all the nation's users could establish "the freest, the easiest, and the least expensive communication between all the different parts of the country, which can be done only by means of the best roads, and the best navigable canals."[50] "By the improvement of our Roads . . . the expense of repairing Roads, the wear and tear of Carriages and Horses, would be essentially diminished," explained another committee in 1810.[51] "Your Committee feel warranted in stating it as their opinion to the House, that there exists no prospect of any new Road, or any sufficient repairs being made, without the assistance of public aid," explained an 1814 committee.[52] The specific arguments about financial efficiency were developed to explain how centralized government would eliminate the problems of local interests, industry monopoly, and technical incompetence in road building.

While gentlemanly administrators pointed to the way turnpikes functioned as monopolies, later writers sought to demonstrate an incompatible conflict of interest in the local administrators and landlords responsible for locally administered roads. Only a centralized board of roads, they argued, would elect administrators freed from the temptation to promote their own interest above the public good.

Eighteenth-century writers established that private turnpikes that acted like monopolies stifled development. Local politicians could and indeed did stop roads that they thought would undercut their own monopolies of markets. Adam Smith reported that London politicians had attempted to squash turnpikes to the suburbs that would have allowed merchants to "sell their grass and corn cheaper in the London market than themselves, and would thereby reduce their rents, and ruin their cultivation."[53] Henry Sacheverell Homer identified "Gentlemen of Property" who stifled roads that would "render the Markets in their Neighbourhood more accessible to distant Farmers" and hence undercut their monopoly of their tenants' wages. As he explained, turnpike commissioners generally took "the Liberty of blocking up the principal Avenues of every other Road which falls into or lands across their's [sic]," turning the whole of the local streetscape into a funnel for their own coffers. "Every Act" for repair, he wrote, "is an

Act also to prevent any of the Roads leading into or across it, be they ever so bad, from receiving the same Remedy."[54] As Homer understood it, then, the system whereby Parliament protected only the sanctity of property rights and the "Monopoly of the Toll" constituted an artificial obstruction to "the Repair of the Roads in general." Neither Smith nor Homer thought that these gentlemen were acting in their own interest; Homer called the suppression of roads "as erroneous as it was selfish." The problem remained that the perceived self-interest of such landlords was the localization, and therefore monopolization, of markets. Leaving road building in local hands meant that in certain significant cases, no roads were built at all. Private transport monopolies stifled development.

After 1790, advocates of centralization developed a broader critique of the problems of turnpike development, characterizing the structure of local government itself as inherently corrupt. Generally drawn from the ranks of local landlords, turnpike trustees tended to be personally interested in the shape of highway development. Trustees frequently used their position to promote their own investments, developing in strips the properties alongside a future right of way and then renting or selling them at a profit once the new turnpike had increased their importance.[55]

Advocates of centralization accumulated evidence that this interest made for arbitrarily crooked roads that obstructed the public's interest in direct transit. Beatson complained that turnpike trustees could not straighten roads because of their interpersonal attachments to the owners of intervening estates: "It would be doing an injustice to human nature to suppose, they can view, with impartial eyes, the fine plantations, the beautiful inclosures, and other improvements, they have made on their estates," he reasoned. "We may as well imagine," he conjectured, "that a doting mother, can coolly and deliberately see an incision made in the skin of a darling child, however much it may be benefited by the operation, as that a country gentleman, can with indifference behold a turnpike road, carried through an inclosure, which he himself has been at the pains and expence of adorning." Beatson complained that the sum of these deviations was a burden to the public: "In other places the traveler and the public, and the poor overloaded horse, are obliged to submit to all the inconvenience, the labour, and the fatigue of ascending and descending the steepest hills, when they might have gone, with the greatest ease and comfort, on a level road."[56] Thomas Grahame, a writer on steam navigation and railways, also believed that the landed interests necessarily distorted the shape of roads:

"The benefits and advantages of direct and well-arranged streets and communications are not appreciated, until it becomes difficult, if not impossible, to obtain them," he explained. He traced the problem to its causes in the squirearchy, the eighteenth-century nexus of interrelated landholders who monopolized political offices in the countryside. Grahame explained "the appropriation of lands by individuals, the formation of inconvenient and faulty lines of communication, and the deep interest, which the various portions of a settled community have, in the maintenance of these faulty lines."[57] Attempting to quantify the potential gain of centralized over local administration, the 1808 committee estimated that variations in turnpike roads increased "the distance . . . about one-seventh part" over the course of a straight line that a centralized administration could create.[58] The 1819 Holyhead Committee accused the trusts of being "very obstinate" against any improvement of their own accord.[59] Highway advocates complained that Britain's roads had not fulfilled their potential service to the public. Alexander Gordon of Culvennen, a local surveyor for the roads of Kirkcudbright in Galloway, was called as a witness by the 1808 committee. He testified, "I have not seen any line of road that may not be made with no greater ascent than one in forty, without lengthening it."[60] Turnpike trustees, as local landlords, held an interest in private property and development that prevented the improvement of the road system as a whole.

Advocates of centralization argued that personal interest and the public interest were in conflict. John Holt complained of the "selfish spirit" that pervaded turnpike administration and referred to frequent cases of turnpike roads to nowhere: "In fact, we may observe in every part of England the *jobbing trade* as it respects turnpike roads, very industriously pursued." He continued, "Personal and local interests frequently supersede a due consideration of general benefit."[61] Robert Beatson could not see how private investors could possibly promote the public good, where "the business of the public . . . is . . . terminated according to the convenience of the strongest party, without any regard to the interests of the community at large."[62] Thomas Grahame concurred: "This class of men, who are both numerous and powerful, are unluckily, too much blinded by their interests and fears."[63] The path roads took was characterized as the result of speculation. John Holt pleaded for a system "founded, *not* on speculations of mere local or private convenience."[64] In 1820, Davies Gilbert, the Oxford mathematician and a member of the Committee on Metropolis Turnpikes, painted trustees and surveyors as unscrupulous profiteers.[65] Wavering between expenditure and debt among turnpike trusts—hitherto standard practice for local turnpikes whose trustees had put up money in the hope of personal enrichment—was redefined as "abusive" and irregular.

The turnpike system's critics accused it of not providing incentives for trustees to make proper financial decisions. The system rather encouraged their absenteeism. As Wright understood it, most trusts started out with some concern for fiscal probity, where "gentlemen of property and respectability" were accustomed to survey "how the money was borrowed, and the income of their tolls were applied, and to take care that every one employed in the concern should do their duty." But over time, nothing prevented them from merely enjoying the benefits of incoming revenue without added labor, as well as obvious possibilities for enrichment: "Their revenue is nearly about six hundred thousand pounds, paid in quarterly, and attended with no expence."[66] In such a system, there was little reason for MPs, themselves the trustees of the district, to lobby for a state system sure to take roads out of their control. Protected by monopolies enforced by Parliament, the turnpike system encouraged passive absenteeism rather than active development. The 1808 committee complained that turnpike trustees "are rather disposed to maintain establishments beneficial to themselves, than to relieve, in an expeditious manner, the public burdens." They pointed to the mortgaged debts that funded turnpike creation and insinuated that the debts were a sign of "a number of abuses" that instead of devoting "the resources of the country" to "useful purposes" rendered them "improvidently wasted." The 1808 committee complained, "In some instances Turnpike Trusts have contracted debts, bearing an interest nearly equal to the amount of their tolls, and when those have been increased, fresh debts are incurred; so that the contributions levied on individuals using the Road become directed to purposes wholly different from their repair." Centralization advocates explained that turnpike trustees did not necessarily reinvest collected sums in better pavement. C. W. Ward, advising the 1808 committee, wrote to its chairman: "It is incumbent on the Legislature, as conservators of the national property, to watch both over their preservation, and the just application of the sums raised from the Public for their repair and improvement." The 1808 Committee on Turnpikes complained that "instead of the roads of the kingdom being made a great national concern," local trusts merely took in "large sums of money... from the public," which they then "expended without adequate responsibility or control."[67] Parnell complained that turnpikes were structured to direct "the spending of the road money as may best promote the interests of [the trustees] themselves or their connections."[68] Through these assertions, profit-motivated management according to the expectations of investors was defined as an inherent sign of private interest in conflict with that of the public.

Robert Beatson complained that "immense sums of money are annually levied for the purpose of making and repairing the highways, yet either from bad management, from party influence, or from chicanery and ignorance of surveyors and contractors, the roads in many places, are not only laid out in the most absurd directions, but are so badly constructed and kept in so wretched a state of repair, that they are almost impassable."[69] The interestedness of turnpike trustees was manifest in the recruitment of bad managers and in haphazard knowledge. The occasional and voluntary correspondence of gentlemanly administrators on the turnpikes seldom specified better means of engineering and managing the roads. As parties explicitly interested in the development of centralized roads appeared, they amassed evidence that expert knowledge would improve roads everywhere, and they encouraged the institution of a centralized authority capable of enforcing standards imposed throughout the kingdom.

The first reliable reports of evidence about the roads were those prepared by John Sinclair in the course of his work to support development in Scotland. Using the painstakingly accumulated testimony of dozens of surveyors from around the country, Sinclair could point to the benefits of gathering expert knowledge in a central location and could recommend that local turnpikes follow the direction of central expertise. These arguments supported Sinclair's claim that local road making was inferior to the potential of centralized building. Sinclair's reports and those of Parnell and the boards of roads after him thus implemented expert knowledge to argue about the inherent superiority of centralized road construction.

Sinclair was a master of gathering evidence, men, and ideas to support political causes. In 1790, he began writing to Church of Scotland ministers, asking them to fill out questionnaires that detailed the lack of roads in their area, the potential for economic growth, and the moral soundness of the local population.[70] These were published as the *Statistical Account of Scotland* between 1791 and 1799. The questions included pointed remarks on the necessity of roads in Scotland, responding to prompts such as these: "What is the state of the roads and bridges in the parish? How were they originally made? How are they kept in repair? Is the statute labour exacted in kind, or commuted? Are there any turnpikes? And what is the general opinion of the advantages of turnpike roads?"[71]

The fact-finding labors of the Scottish clergy continued under parliamentary sanction after 1794, when Sinclair arranged, with the help of Sir Joseph Banks and Arthur Young, for the newly formed Board of Agriculture to begin collecting information on the state of agricultural practice in the entire island "to ascertain the real situation of the country, and the means of its improvement" by virtue of systematizing a "vast mass of information and experimental knowledge."[72] Scottish integration with the British economy remained a major focus of Sinclair's work on the board. As the board's first president, Sinclair loaded the ranks of official correspondents with his fellow Scots, who, not surprisingly, were prone to suggest the importance of roads leading northward. Many of these correspondents were the same Scottish clergy whose information Sinclair had used in his *Statistical Account of Scotland*, men Sinclair referred to as his "statistical missionaries," willing to travel to the periphery of the island on fact-finding missions.[73] Scotland's interests were well and fully represented, then, in the commentaries with which the Board of Agriculture advised Parliament.

The Board of Agriculture siphoned money to these largely Scottish agrarian reformers, who pumped the bookstores full of their tracts on improving the economy. All included exhortations for the roads, many of an evangelical frame, pointing explicitly to the failures of turnpikes and local administration.

The writers blazed through the traditional objections that had riddled eighteenth-century thought about the roads, affirming that property disputes would no longer be a problem. Where English writers were made pessimistic by experience, Scottish writers alone could confidently assert that "a turnpike road cannot be carried through . . . without the proprietor's consent, or an act of parliament." On the basis of this unearned trust in law and order, Robert Beatson could predict a world of centrally planned highways and just reimbursement where "every man is allowed to enjoy unmolested whatever he is possessed of."[74]

Much of the rest of the board's correspondence consisted of purely practical suggestions about the technological improvement of roads. Were roads and canals built to expert standards, it argued, the general increase in revenue would enrich the nation enormously. Following the pattern pioneered by the Highland Society, the Board of Agriculture correspondents exhaustively detailed the examples around the nation from which they proposed an ideal model. John Holt surveyed international examples for the best designs for guideposts, guard posts, snowplows, and milestones.[75] Robert Beatson turned professional surveying knowledge to the purpose of explaining how to discern and render most accurately the shortest, flattest, and most foundationally secure route between two points.[76] He explained the successful experiments with concave roads run by Mr. Bakewell of Measham in Leicestershire and Mr. Wilkes of Bredon, where rain pouring through a center sewer automatically washed the surface clean. He summarized the techniques developed by John Metcalf for preparing roads on the sides of hills and laying a foundation through a swamp. He recommended a machine for breaking gravel and a harrow for smoothing roads, invented in 1786 by Mr. Harriott of Great Stambridge in Essex. He touted the horse-drawn railroad made through a peat bog near Manchester by Mr. Wakefield, which could be "drawn by one horse over a moss, where, a few months before, even a dog could hardly venture without the risk of being *swamped*."[77] Such detailed accounts of local success, foregrounded by Sinclair's encyclopedic collection of case studies, lent credibility to the proposal that expertise existed, and that a centralized institution could coordinate its flow to the benefit of all.

The work of the highway lobby rested on the overwhelming masses of details assembled to support its arguments. This was a skill that Sinclair had cultivated during his work with the Statistical Survey and Board of Agriculture. From the beginning of these proposals, Sinclair outlined expert knowledge as the basis for government reform. He laid out the history, basis, and examples of statistical surveying for governmental purposes in an address to Pitt printed in 1793 as *Specimens of Statistical Reports*, representing his work as consistent with "the anxious attention to facts" rather than the fancies of "visionary theory."[78] He constructed the project as an extension of traditional statecraft, conducted on a rational basis: "Real statesmen, and true patriots, can no longer be satisfied with partial and defective views of the situation of a country, are now anxious to ascertain the real state of its agriculture, its manufactures, and its commerce."[79] Indeed, Sinclair is remembered in history not as a great statesman or as a pioneer of roadways, but as the man who introduced the term "statistical" to the English language, and who pioneered many of the first exhaustive studies of the public economy.[80] These talents struck contemporaries by their novelty, as well as by their almost comic modesty. The London press joked that Sinclair was hardly to be praised for "expansive comprehension" and "eagle-eyed intuition" or even "those trembling sensibilities of sympathy and of passion." Compared with even mediocre masters of rhetoric, Sinclair was unreadable. This particular editor stooped to sarcasm: "His are not the powers which delight to ride in the whirlwind, and to controul the storm." Sinclair had rather "descended, to dwell in the regions of minute details, without having, first, duly cultivated his powers of abstraction and generalization," which would have presumably made the accounts digestible.[81] Sinclair had pioneered, in a sense, the rhetoric of the modern political report and technical manual: a collection of facts in no sense meant to cajole, inspire, or even illuminate.

But Sinclair's encyclopedic tomes remained convincing by dint of their sheer bulk. Even while he ridiculed the Scot for his lack of rhetorical drama, the editor nevertheless conceded that the swarm of facts had great power, capable of conceiving "a magnificent project; ... in spite of every obstacle," and carrying it "fairly and effectually into execution." The magazines mirrored that Sinclair had almost single-handedly inaugurated a new government institution in the Board of Agriculture, and they marveled at how, in the process, he had stimulated an original "science of agriculture, which, before his time, had scarcely an existence."[82] What Sinclair had essentially invented a new kind of political tool that was based on drowning his enemies with an overwhelming onslaught of quantified facts. Communication with a wide network of committed correspondents meant information, and information meant political power.

Writers for the Board of Agriculture argued that their mass of specific evidence could be properly enforced only by a centralized board of roads. Robert Beatson urged Parliament to create a standing central authority with the power of "issuing orders" to the trusts "for making and repairing" roads and securing enough "money for that purpose." As Beatson imagined it, this body would keep roads out of the hands of investors and safely in the hands of engineers: "Under this Board should be appointed, the most able surveyors and inspectors." Centralizers envisioned an institution that would be truly national in affiliation and interest and thus eliminate the conflicts of local personality. To keep its engineers free from local interest, it would follow the pattern established for excise officers of circulating from county to county, so as to better develop their ability of discerning a national rather than a local standard of perfection in road building: "To each of [them] should be allotted a certain county or number of counties; and they might be changed annually, or triennially, from one district to another, that they may the more generally know the best practices followed in different places, and be the less liable to form intimacies or partialities."[83] As formulated by highway advocates, centralized road building would eliminate the conflict of interest inherent in local politics. Only a centralized authority, reasoned highway advocates, could ensure the disinterested management of roads throughout the entire kingdom.

In Parliament, highway advocates developed the resources of innumerable kinds of experts, catering to differing visions of how better legislation could be implemented. At first, these committees were content to raise limited questions about the design of wheels and carriages, suggesting a willingness to entertain the pattern of broad-wheel legislation that characterized eighteenth-century government regulation of the roads through restrictions on individual users. In 1806, 1808, and 1809, the committee interviewed dozens of carriers, stagecoach makers, wheelwrights, axle smiths, and local surveyors about the best possible regulations of individual vehicles, the problems of local enforcement, and the best hopes for maintaining the roads.[84] After compiling volumes of reports, the committee switched to press the issue of centralization. "In fulfilling the duty imposed upon them," it wrote, it had reviewed the carriage-building habits of the nation, "yet they are led to believe that no system regarding either roads or carriages will be complete, unless some Public Institution or Department of Government is entrusted with the power of enforcing the due execution of those Laws." Broad-wheel legislation, under the authority of local legislators, was beyond any system of enforcement and vulnerable to local corruption, local variance, and the interest of the trustees rather than the interest of the traveler. In place of broad-wheel legislation, the Committee on Broad Wheels recommended a centralized authority, "which the wisdom of Parliament, aware of the infinite importance of the subject, may from time to time judge it essential to enact.[85] Treatises on road making by Macadam and Telford reduced road making in very different climates to systems whose costs were to be quantified and predicted.[86] These figures served to remove the distinction between buildable lowland and unbuildable highland roads and turned both into questions of current investment and long-term benefit. Numbers and quantitative evidence, Telford argued, would protect the interest of the nation rather than allowing it to fall into the corrupting influence of personal relationships and the desire for riches. Strategically picked experts from a variety of professions thus offered the committees on highways evidence that always pointed to the importance of centralized expertise in general.

Road-building committees advertised the work of these experts as a feature of their promise. Telford offered his reams of maps, statistics, accounts, and reports of the process of vetting contractors in the hope that "the more they are investigated, the more important they will appear."[87] The 1819 Holyhead Road Committee related that it had been "proved in the course of this Session" as fact "by Mr. McAdam, that a very bad road may be made a very good one." Macadam's system, it explained, demonstrated that even routes through difficult territory could be provided with a solid foundation "at the ordinary expense which is incurred in repairing all the heavy gravel roads about London, by attention to cleaning the gravel and breaking the larger stones."[88] Telford explained that by reviewing these designs, "the Public will become more fully convinced, that the general Interests of the British Empire are extensively connected with the several Improvements which are mentioned in Your Lordships Instructions."[89] Committees identified specific evidence in maps, charts, and figures of roads that could be improved only by expert engineering. The 1808 Turnpike Committee reckoned that the roads of every city in Britain stood to be improved, and it considered evidence that their current routes, having "been formed by gradually widening the paths made in early society, partake of all the deviations to which those paths were subject."[90] It too reasoned that centralizing roads was the only solution, and that the highways "cannot be brought to that state of perfection of which they are capable, without some attention on the part of the Legislature, nor by Committees of the House, occasionally appointed, however zealous in the cause."[91]

Local government appeared in these debates as the single feature that most inhibited better engineering. Robert Beatson, writing for the Board of Agriculture in 1793, complained that local governments tended to put power in the hands of local surveyors; trusts tended to "trust the inspection and whole management and direction of the roads to some ignorant or pretended surveyor; who, almost to a certainty, will impose upon them, especially if he is empowered to settle with contractors." Beatson illustrated turnpike incompetence by citing the many examples of meandering routes in the nation, where "roads are directed in the most irregular zig-zag manner, through a level part of the country, where they ought evidently to have gone straight forward."⁹² The 1808 committee promised that its care would eliminate such irrational traces of the past: "Those deviations were made originally to avoid obstructions no longer in existence; in some places, ascending hills to avoid woods or marshy tracts, now become open and solid; in others, inclining from a direct line for the opportunity of fords, now rendered unnecessary by the erection of bridges."⁹³

In Parliament, advocates of highway centralization employed similar arguments, promoting centralized expertise against local government. Local governments, they argued, appointed officials qualified by neither knowledge nor impartiality for the weighty tasks before them: "The practice is to make almost everyone a trustee, residing in the vicinity of a road, who is an opulent farmer or tradesman, as well as all the nobility and persons of large landed property," explained Henry Parnell. Such large, participatory bodies stifled the wiser voice of technological expertise: "The result of this practice is, that in every set of trustees there are to be found persons who do not possess a single qualification for the office." Worse, power made these people arrogant, capable of "opposing their superiors in ability and integrity, when valuable improvements are under consideration."⁹⁴ Turnpike trusts, the 1808 committee suggested, simply did not have the technology to discern the true course if they wished to pursue it. Employing local surveyors, it insinuated, they often lacked the tools of modern geometry, and as a result, their roads "in all cases pursu[ed] an apparently straight course, by surmounting ascents, the arch of which is at least equal in length to the level line formed by circling their base."⁹⁵ Ignoring the fact that road-straightening legal tactics had originally evolved at the parish level, centralizers condemned the trusts where road straightening had not yet happened and associated the lack of straight roads in many trusts with local incompetence.

In the climate of local interest, Parnell reasoned, talent and competence, whenever they appeared, were subject to abuse, injury, and burnout. If "one trustee, more intelligent and more public spirited than the rest," should make a proposal founded on expert knowledge, "a measure in every way right and properly to be adopted, his ability to give advice is questioned, his presumption condemned, his motives suspected." No intelligent engineer, Parnell thought, would want to put himself in the position of demonstrating his evidence before an adverse audience composed of mixed and diverse aims. Instead, "frequently experiencing opposition and defeat at the hands of the least worthy of their associates, they are annoyed by the noise, and language with which the discussions are carried on, and feel themselves placed in a situation in which they are exposed to insult and ill usage." The very process of engaging a participatory, diverse, and democratic audience, Parnell reasoned, alienated the most creative and technologically competent persons involved in road building: "Intelligent and public spirited trustees become disgusted, and cease to attend meetings." Democratic meetings on the local level were simply too difficult for engineers and visionaries, who would inevitably be defeated by local demagoguery. Parnell argued that expert engineers never won on the local level: "As every such measure will, almost always, have the effect of defeating some private object, it is commonly met either by direct rejection, or some indirect contrivance for getting rid of it."[96] Centralizers had amassed evidence about their own plans, backed up by surveys, maps, and data, which allowed them to paint local government as a scene where the effectual implementation of expertise was impossible.

National Integration

Eighteenth-century critics of turnpikes pointed out how the varying patterns of regulation across different localities and regions punished the traveler. John Scott, a trustee of the Chestnut, Wadesmill, and Watton Turnpike in Hertford, complained that the regulations on highway use as a whole were a matter of "disjointed clauses," causing "no small Degree of Perplexity." The traveler was bound to confuse permissive regulation and strict statutes on his journeys: "In one Place he met with general positive directions, which he depended on as authentic rules of conduct; till he perceived, that in another, they were counteracted by particular exceptions; and not unfrequently, he saw subjects, closely allied in their nature, removed almost as far from each other as the utmost limits of the act would permit."[97] Frustration with corruption, uneven enforcement, and the confusing letter of the law conspired to make those legislators who dreamed of an easy form of highway regulation desperate. By 1770, it was clear to most that the regulations were too confusing, too contradictory, too complicated, and too ill administered to have much effect. Reform measures, tolerated and discussed for two decades, were finally met with apathy. The result was that the old regulations on weight were allowed to expire in 1765 and were replaced by new fines that applied only to weights above five tons.[98] As a preliminary and gentle step in the direction of mutually improved roads, broad-wheel legislation was a splendid failure. It highlighted divisions by section and profession and the unwillingness of one region to submit to another's interest. John Wright of Chelsea explained that turnpike gates and weighing machines were "erected at an immense expence and then farmed out." He concluded that corruption was endemic. The machines were "a nuisance on the roads, and ought to be abolished."[99] Local road administrations, with their arbitrary sets of rules, resulted in an irrationally expensive system characterized by difficult travel.

Turnpikes and local governments were incapable of making roads that spread wealth beyond a particular region. As economic ventures, they invested in those areas of the country where trade was most intensive. Staggered and occasional turnpikes appeared in Scotland, Wales, and Ireland only at the beginning of the nineteenth century. No turnpikes existed in more remote places like the Highlands of Scotland, where resources for travelers were scarce.[100] Adam Smith explained that roads had the unique power to open up competition in new regions. Roads "put the remote parts of the country more nearly upon a level with those in the neighborhood of the town."[101] Such critiques of the turnpike system made clear that centralized government could press national integration where local governments could not.

Once again, the best-bolstered arguments were those formulated around the ambitious efforts of Sir John Sinclair. Sinclair's personal interest in national integration and the fierceness with which he fought for the roads were both motivated by an intense personal investment in the development of Scottish industry. Correspondence with Adam Smith had convinced Sinclair that Scotland could reach economic parity with England only if England paid for the development of highways that would allow Scots to participate in Britain's booming economy. Writing in his *History of the Public Revenue* (1785), Sinclair had outlined the connections between highways and the development of the national periphery. For Sinclair, national assimilation could most rationally be brought about by the centralized building of infrastructure and the promotion of new markets. He explained the goal, equal market access for all: "The remotest parts of a kingdom are thus gradually brought to be nearly as valuable and important as those situated in the neighborhood of the metropolis." The introduction of trade would gradually multiply into the development of local industry. Highways, Sinclair observed, could bring about the development of the Scottish economy, ensuring the health of Scottish peasants, the education of Scottish youth, and the promotion of sustainable Scottish industries within a larger market. These industries would then enrich England itself, transforming the separate kingdoms at last into "one firm and compacted body" defined by its mutually enriching markets.[102] A nation united by highways would thrive, he promised, far more than separate kingdoms impoverished by their own closed markets. Under Sinclair's direction, experts commissioned by the Board of Agriculture made similar arguments about a nation connected by highways. Roads, Beatson wrote, "are as the veins and arteries to the human body." He explained, "If this circulation is by any means checked or obstructed, even in the remotest part, that part soon becomes useless, and sinks into decay, and in some degree is felt throughout the whole body."[103] John Holt pleaded for legislation that would benefit not "particular towns, districts, or even counties," but rather "the more extended considerations of general intercourse and common benefit."[104] Traffic alone could ensure that all of the nation's parts would work in harmonious concord to the mutual enrichment of all. Such a unified, market-driven, and equal nation offered the only terms on which Scots could imagine themselves operating in England as equals.

Sinclair proceeded by allying himself to those individuals most likely to have the interest and ability to help the cause of Scottish development. In 1784, he was a founding member of the Highland Society, whose Edinburgh and London branches consisted of landlords and investors devoted to promoting Scottish culture and trade. According to his daughter, he immediately proposed to them to obtain a grant of £50,000 from Parliament to build roads "throughout the ultra-northern counties, where the drivers of cattle had to swim with their droves across the rivers when taking them to market."[105] Scottish nationalists in places of influence, the members of the society were ideally situated to help Sinclair secure his vision. The rules of the society strictly limited membership to "Natives of the Highlands of Scotland, the Sons of Highlanders, Proprietors of Lands in the Highlands, those who have done some signal Service to that part of the Kingdom, Officers of Highland Corps, and the Husbands of Highland Ladies" and thus bound the institution to a strictly ethnic and nationalist interest.[106] Its first activities, like Sinclair's own, addressed a nationalism built on Gaelic language and literature.[107] In 1786, however, the society began commissioning excisemen in Scotland to count the number of fishing vessels in the Scottish ports, conduct their own estimates of the possible expansion of the economy, and submit their own reports to Parliament's Committee on Fisheries, arguing for the founding and promotion of new villages on Scotland's northern shores.[108] The society offered prizes for the best designs for these plans, and the plans became evidence in Parliament of the solid foundations and merit on which their arguments rested.

The Highland Society came together at a moment when regions were already becoming political and economic actors, vying with one another for parliamentary control. Over the course of the century, Britain's economic geography had taken on a new form, supplementing political controversy with a new array of potential actors. Regional economies had developed and specialized, with a budding industrial center in the Midlands and a new agricultural belt across the north.[109] The Scottish economy began developing in its own right after 1745. After union in 1800, Ireland was added as a player. Financial interests and migration fixed London as the preeminent city of the nation between 1800 and 1840.[110] Battles over taxation in the 1760s had finally splintered the old court-country rivalry of political factions and had created a parliamentary divide informed by regional and economic interests instead.[111]

The Highland Society stood out in its activity for its talent at linking regional interests in Parliament to moral questions like poverty. Members of the Highland Society pointed to Scottish poverty and Scotland's removal, geographically and institutionally, from the sources of capital capable of paving new highways through private or local investment. In Scotland, the mountains and moors housed only a few poor shepherds and cattlemen, incapable of funding the entire enterprise themselves. The society reported to Parliament in 1813 that necessary roads might well "traverse certain districts capable of no agricultural improvement, and inhabited only by the Shepherd and his Dog."[112] Parliamentary committees gathered to address the findings, and argued that roads should be provided by those with the greatest means for those with the greatest need. The Committee on the Carlisle and Glasgow Road gave the example of Rannoch Road, which passed through "a most rugged Country from Fort William," with sparse inhabitants incapable of paying for the road.[113] Local Scots in their poverty were unable to pay for the envisioned highway: "It passes through a mountainous, pasturage, and thinly peopled country, to whose immediate interest a more improved communication is of little comparative importance."[114] Nationally administered infrastructure could develop regions that did not have the interior resources to develop on their own.

Roads to distant provinces, advocates promised, would lead to their development. Dreaming of riches from Scottish roads, Hawkins Browne, the industrialist MP from Shropshire, imagined that "by opening the communications with roads and bridges, not only the conveyance to and from the fisheries would be facilitated, but the business of agriculture promoted."[115] J. F. Erskine explained the general rise in rents that he expected if highways were paved in Scotland: "It may be fairly reckoned that they increase the rents of those lands from 2s 6d to 10s per acre, according to the goodness of the ground, the state of the road before the improvement was made, or other local circumstances."[116]

The development of provinces would thus lead to general wealth by increasing commodity competition, driving down rents, and thus producing affluence all around. Roads, Sinclair explained, increased the circulation of information necessary for other kinds of commerce: "The health of the inhabitants is also preserved by traveling about, in surveying and visiting their own country; and improvements and information of every kind are more rapidly and more easily communicated."[117] The 1813 committee too explained how England, as well as Scotland, would benefit from the existence of connective highways: "By the conveyance of these Roads, the South receives annually a great supply of *Sheep*, Cattle, Wool, and other articles from the Northern and Highland Districts."[118] It continued, "By the improvement of our Roads, every branch of our Agricultural, Commercial, and Manufacturing Industry, would be materially benefited— Every article brought to market, would be diminished in price."[119] "The advantages of this Road are more of a general than a local nature," urged the 1815 committee on the proposed Parliament-built highway between Carlisle and Glasgow, which would pass through many uninhabited wastelands. Such arguments, straight out of Adam Smith, seemed to confirm that nationally built roads across the kingdoms would increase the wealth of both kingdoms. "The importance of Land-carriage, to the prosperity of a country, need not be dwelt upon," argued the 1808 committee for road reform.[120] Thomas Telford, speaking to the 1802 committee, urged that roads that reached into "the remote Points" of the nation would offer new frontiers for economic exploitation, and it described Scotland as "the best Field for useful Exertion to the present Seats of Capital and Industry."[121] John Holt, the Board of Agriculture correspondent for Lancashire, described roads as the basis for "public convenience and prosperity."[122] In Parliament, advocates promoted highways as a formula for a nation bound together by trade and wealth.

Roads also promised to soothe racial tensions on Britain's uncivilized fringe. Telford explained in an early report his hopes how roads would change the Scots: "They would by this Means be accustomed to Labour, they would acquire some Capital, and the Foundations would be laid for future Employments."[123] In 1803, Telford explained to a group of assembled MPs, "The Want of farther Roads and Communications in the Highlands, has hitherto proved the greatest Obstacle to the Introduction of useful Industry there." It advised that "regular and easy Communication ... from one Part of the Country to another" was the solution.[124] Roads could transform lazy and unindustrious people into hardworking Britons. Finally, development would generate political consensus among formerly polarized geographies. "By such a communication between the capital and the country," Sinclair argued, "the whole society becomes, in a manner, one firm and compacted body, impressed with the same ideas, actuated by the same principles, speaking the same language, animated by the same spirit, and in every respect resembling the fellow-citizens of the same town."[125]

Roads would promote national safety by protecting the nation's supply of soldiers for the army. Highway advocates promised that roads would boost the economies of the ailing Celtic fringe and thus help preserve one of the traditional sources that "furnish officers for our fleets and armies."[126] Hawkins Browne, an English industrialist on the Highland Roads and Bridges Committee, pointed to suggestions that roads would stanch Scottish emigration and reasoned that highway building was "an extraordinary Case" "where such a numerous Body of the People are deeply interested" as to justify Parliament in considering highway building "the Duty of Government" and therefore "departing a little from the Maxims of general Policy" and taking a course of centralization.[127] Telford declared one of his purposes in the 1802 survey of possible roads in the Highlands to be the ascertaining of "the Causes of Emigration and the Means of preventing it."[128] Telford warned that if the government did not intervene, "our Armies and Navies will then be no more." He explained "that emigrations have already taken place from various parts of the Highlands." Telford's sources estimated about three thousand in the preceding year, "and, if I am rightly informed, Three Times that Number are preparing to leave the Country in the present year." He blamed this situation on the fact that poor farmers who could "see no Mode of Employment whereby they can earn a Subsistence in their own Country" had been "deceived by artful Persons" into thinking that a better future waited for them in America.[129] Hawkins Browne was persuaded that "the country being rendered thus tolerable to its hardy and industrious inhabitants, they would be no longer induced to emigrate."[130] A similar argument was later made for Wales during the debates on the Holyhead Road. In 1816, Mr. Jones favored "some improvement in the road acts" to cushion the Welsh peasantry who had "contributed to the military strength of the empire, and at a critical moment had rushed forth to defend themselves from foreign invasion."[131] Roads, by economically integrating the peoples on the nation's geographic periphery, would provide Britain with a constant stream of bodies for the purposes of national defense.

Centralizers played on British fears and self-preservation with reference not only to international threats but also to everyday dangers at home. One witness recalled a recent crash where "one woman had her leg and thigh broken, one man had both his arms broken, a third his shoulder dislocated, and all the rest were very much bruised and wounded, some having their ribs broken, and otherwise much maimed."[132] The Holyhead Road Committee described dozens of cases of broken legs, terrified passengers, horses collapsing from exhaustion, broken coach poles, and bodies hurled off the tops of coaches.[133] Committee reports agreed that these disasters could be solved by slight improvements in engineering. In 1802, Telford warned "that Accidents frequently happen from the want of a Bridge over the River Tay at Dunkeld in Perthshire."[134] In 1809, arguing for the building of new roads and bridges, he described the current bridge over the River Eden at Carlisle, only twelve feet wide, as so narrow that "no two carriages can pass each other without danger of being crushed to pieces, or pushed over the parapets: it is even dangerous for horses or foot passengers to pass a carriage or cart." He urged that it was "very urgent that a new bridge, of proper dimensions, should be erected at this place."[135] An 1808 witness to one of the worst accidents explained, "These accidents happen more frequently than the Public is informed of." He concluded, "Some Parliamentary interference is certainly necessary."[136] John Proctor, testifying to accidents on the Holyhead Road that could be corrected by better highway maintenance, reasoned from the extent of disasters that "it is indeed high time that the Legislature should interfere, and put a stop to such malpractices."[137]

Vivid examples of the spectacle of bodies in pain made a visceral argument for increased highway expenditure. The highway committees benefited from a wealth of hard data available from the Post Office, which kept excellent statistics that the Holyhead Road Committee cited as part of the case for government intervention. The 1810 committee laid out the influence of highways on the public mail transports, "which were delayed in less than 85 days" "no less than 71 times, varying from under one to five hours."[138] The appendix included extracts of different time bills showing the average of the Holyhead Mails, a table of accidents on the mail coaches between Shrewsbury and Holyhead and Holyhead and Chester in 1809 and 1810.[139] Post Office data filled in a terrifying picture of a nation at risk of maiming, death, and broken limbs. "Accidents are perpetually happening," explained the committee members, and "scarce a week passes without some of those Carriages breaking down, and often killing the unfortunate Passengers who have trusted themselves to that mode of conveyance." They listed examples of a "Coachman . . . killed," of passengers "dreadfully maimed," and other cases of "material injury" from the Croyden, Portsmouth, Bath, and Liverpool coaches in the previous six months.[140] In numerous instances, the road surface or the "decayed and ruinous situation of one of the Bridges" was to blame.[141] "In short," summed up one committee, "the instances are innumerable."[142]

We know that these lurid stories and scenes took hold in the popular imagination from the popular descriptions of perils on the road. In 1798–1799, 1806, and 1814, Britain suffered worse snowstorms than it had seen for a generation.[143] The blizzards were deep enough to cause lost mail coaches to go missing for days. A postboy was dug out of the snow, frozen to death; and passengers died, frozen in coaches or lost from the lack of landmarks. Ten- to twenty-foot drifts stopped all the roads out of London.[144] The snowstorm of 1814 created a national sensation and opened up public debate in newspapers and magazines about whether Parliament should take responsibility for an amenity on which all of Britain depended. The *Times* clamored for better pavement on the basis of these deaths in a series of lengthy articles.[145] Painters and lithographers began to capture the image of a nation at risk on the road in a state of mutual imperilment.[146] The popular lithographer Henry Thomas Alken produced a series of nine pencil and chalk

Henry Alken, *The Holyhead and Chester Mails*, 1837.

drawings titled *Stage Coach in a Snowstorm*, which included the Holyhead Mail trapped and horses buried up to their necks by drifts of snow.[147]

Advocates of highway centralization argued that expert administration of labor would simultaneously produce better roads and eliminate national poverty. Using road building to accommodate soldiers would solve a crucial problem of reforming the military state: creating a domestic, peacetime outlet for the military that would actually promote economic growth. John Sinclair, citing the example of military roads in Scotland as a precedent for peacetime application, agreed: "The military cannot be better employed, than in carrying on public works, of so useful a description as roads, canals, &c."[148] Robert Southey agreed: "The evil consequences of the idle hours which hang upon the soldiers' hands are sufficiently understood," he wrote. "Would it not be well to follow the example of the Romans, and employ them in public works?"[149]

Others extended the provision of welfare from soldiers to the nation's poor in general. The Holyhead Road Committee argued that a national scheme for road management could afford "immediate relief . . . to the Poor, from one end of the line to the other." It even boasted that no form of relief could cause "greater advantage to the Public" than these road-building schemes, "because whatever sum of Money might now be expended amongst the Poor, would command a great deal more Labour than the same sum could do" in other circumstances that did not carry the double benefit of poor relief used for infrastructure.[150]

In order for road construction to accommodate these political needs, the formula for road construction had to be refined. Earlier modes of road repair, including the first models encouraged by Telford, planned for smooth road surfaces handled by a few skilled laborers. Fine craftsmanship was required for the carving and setting of heavy flagstones seven inches thick that had to sit exactly square against one another. Against this model, Macadam proposed a model of management of cheap, unskilled labor on gravel roads. Macadam touted his rules throughout his writings as a systematic way of dealing with the out-of-work poor by putting them to work on the roads. He designed the basic outline of his system to do away with those skilled positions of professional pavers that Telford's system of fitted flagstones required and to replace expensive craftsmanship by a few with basic stone breaking by the many. His model was cheered by road advocates, who greeted it as a solution to social problems. Thomas Hughes, in his road manual of 1838 explained the new "systematic mode of improving the road" as characterized by the "principal ingredient" of "labor."[151] Others pointed to the "great advantage attending Mr. MacAdam's model of road-making," the discipline of "human labour," and to the changes whereby the majority of highway budgets, previously dedicated to the rental of horses, were now dedicated to paying for more efficient and productive forms of human labor.[152]

Macadamized pavement, or broken-stone pavement, called for less craftsmanship and more labor. Above a lower level of gravel was a top layer of dressed paving stones, cemented with mortar and beaten into place by a paver with a fourteen-pound "wooden maul" of beech or elm.[153] Every stone on the bed of a macadam road was "broken into pieces as nearly cubical as possible, so that the largest piece, in its longest dimensions, may pass through a ring of two inches and a half inside diameter." The work of this project was broken down into individual, assembly-line tasks, assigned to different corps of laborers: the carting of rocks to the site, the sorting out

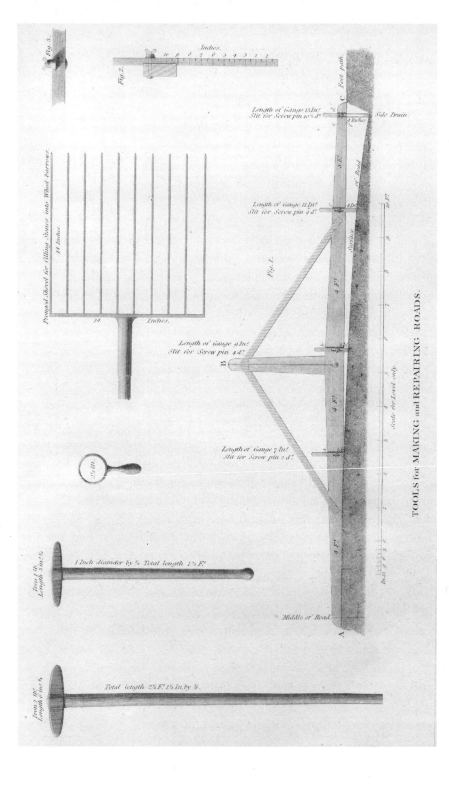

of gravel stones too large for the roadway, the breaking of large stones down to useful sizes, and the laying of stone in specific dimensions. Divided into basic components, road making became a reproducible, mass activity that could be executed by unskilled laborers with no professional training in the carving of stone. The road then had to be maintained on a regular basis by "scraping" the top gravel, a process that involved removing feces and dirt and washing the whole. It was for this reason that the annual maintenance rate of the Blackfriars Bridge went from £100 to £1,500 after macadamization; macadam was a labor-intensive roadway. Intensive labor continued not only when the road was laid but also throughout its life.

The management of cadres of unskilled labor was made possible by the adoption of new techniques for the oversight of workmen.[154] The engineers specified that gravel chunks six to eight ounces in weight were ideal. They devised different tools to ensure the uniformity of the gravel. In some places, "a pair of scales and a six-ounce weight" were handed out to laborers. In others, they were given an "iron ring two inches and a half attached to a handle, through which every stone must be small enough to pass." In other places, the workmen were equipped with an iron ring of fixed diameter to ensure regularity in the stones' sizes. Every morsel of gravel on the British roadway was passed through these measuring instruments, stone by stone. Another set of workmen was assigned to breaking stones with hammers: "The uppermost stones then had their irregularities . . . broken off by the hammer, and all the interstices . . . filled with stone chips firmly wedged or packed by hand with a light hammer."[155]

Both the sorting and the breaking of stones called for specific tools that ensured a homogeneous management of the stones on the road surface with a minimum of external oversight. New tools for sorting rocks and grading curves helped the rank-and-file workman conform to the designs for proper drainage invented by the engineers.[156] Primitive tools then merely made the bodies of labor conform to the needs of design, thus rendering the discipline and use of human labor more productive and efficient. Telford provided diagrams with the exact dimensions of the side drains, slope, fences, and gravel. In his 1820 report, he set out general rules

---

"Tools for Making and Repairing Roads," in Thomas Telford, "Report," *House of Commons Sessional Papers*, 1820, Coms. for Improvement of Holyhead Roads, on State of Road between London and Shrewsbury (126), p. 57.

for materials, differentiating stones from gravel, depending on the availability of both. In the case of stone, each "should be so broken that it may, in its largest dimensions, pass through a ring of 2½ inches in diameter." He instructed surveyors and workmen that "hammers with slender handles, light and well steeled," must be made for the specific purpose of the task. This work ought always to be done by measure, either at the quarries or in proper recesses, made for the purpose on the sides of the road."[157] The experience of road making would be standardized down to the tools. The tailoring of tools and management to unskilled labor made road building into an instrument of mass labor management.

As a result of the tools, the engineers and surveyors were left with only basic tasks of management and oversight that were easily applied to a large number of workers. The men setting these stones were overseen by an "inspector" who watched "all the operations" as they passed and then walked "over the pavement" when it was completed to try with his feet whether it was "firmly fixed." Macadam designed the system to ensure that these managers retained diligent control over wages as well as activity. In advance of any work being done, the surveyor in charge would lay out detailed schedules that recorded the "specifications of the work, of every kind, that is to be performed in a given time," and then would mark out who had performed the work before any pay was delivered.[158] These practices ensured that pay was matched to the particular work of the tasks accomplished, and that unskilled labor was rewarded with set piecework. Macadam's "General Rules" warned parliamentary surveyors that "all labour by day wages" was out of the question. Road making would operate as an extension of the disciplined, industry-enforcing mechanisms of the state, exactly matching money to labor by surveilling and punishing the idle.

As a result of these plans, highway advocates could further boost their plans for centralized rule. Modern surveyors began uniformly to adopt such measures, consulting manuals by both civil engineers, like Thomas Telford and John Loudon Macadam, and politicians and landholders, like Henry Parnell and Richard Lovell Edgeworth, whose long exposure to problems of management in the role of highway advocates made them as well acquainted with the principles of foundation maintenance as any engineer or surveyor.[159] Member of Parliament Thomas Mostyn testified to the "hundred of poor honest fellows endeavouring to get employment, and their families literally starving." It was thus, centralizers argued, that centralized highway labor offered a means of economically integrating the nation's poor. At a professional meeting of engineers, Telford explained this viewpoint: "I consider these improvements some of the greatest blessings ever conferred on any country." He explained: "£200,000 expended in fifteen years has changed the moral habits of the great working class for the better, and has advanced the country at least one hundred years."[160]

The highway reforms succeeded in both rebuilding the nation's major corridors and restructuring the entire local fabric of highway management. Highway partisans saw the committees for Scottish and Irish roads as precursors to one, centralized authority with control over Britain's roads as a whole. "From all the information that has come before Your Committee," wrote the Holyhead Road commissioners, "they feel quite convinced that this principle of consolidating the Trusts is not only wanting upon the Holyhead Road, but upon every great line of road in the United Kingdom; and that until it is adopted by parliament, the system of small Districts and large bodies of Trustees must prove an insurmountable impediment in the way of removing the great defects which are now to be found on all the Turnpike Roads."[161]

Parnell submitted a series of bills with the intention of instituting a new board of roads that would have power over the entire road network. These reforms had limited success, however. After seven revisions, the first round of recommended changes passed in 1810.[162] It included new regulations of weights, new clauses to enforce the provision of footpaths alongside the highway, and the enforcement of cylindrical rather than conical wheels. Further piecemeal reforms enforced a vision of mutual responsibility at a national level. By 1820, new forms of traffic regulation laid down the first laws against "reckless speed" and punished speeding post-coach drivers.[163]

New legislation put new limits on who could be a trustee and divorced local landholders interested in particular improvements from membership in the boards responsible for roads. Charles Dupin, the French observer of British infrastructure, reported approvingly, "It is required that a trustee should abstain from exercising his functions whenever his own private interests are concerned; and that no individual keeping a publichouse, inn, tavern, &c. can become a trustee, or possess any place deriving a salary from the trust."[164] The 1835 act put new limitations on the interestedness of surveyors and contractors, forbidding surveyors who determined the need for repairs from partnering with contractors who made the repairs.[165] Circumscribing the interestedness of turnpike trustees, calling for certain actions on their part, and delimiting their actions in other areas, the new legislation essentially turned trustees into government bureaucrats.

Rebuilding Britain

By 1835, then, the administrative transformation of Britain's road system was complete. Parliament had become the builder of highways, the architect of new engineering and urban design projects on an unprecedented scale, and the centralized coordinator of a well-managed, national body of highway bureaucrats scattered around the country. Between 1803 and 1835, Britain's Treasury financed the building of roughly seventeen hundred miles of roads that radiated from London to the north and the west across expensive embankments and suspension bridges. It controlled an equivalent of 10 percent of the turnpike system, and plans were being introduced to take control of the remaining turnpikes and parish roads. Several of Britain's major corridors had been constructed from the public purse, with maintenance expenses deferred by public expenditure, managed by funds that reported straight to Parliament. The Highland Roads, the Liverpool Road, the Carlisle Road, and the Holyhead Road formed a northward-branching tree of nationally built corridors that with nationally collected moneys tied Scotland closer to the metropolis. Between 1800 and 1835, the political work of highway partisans transformed the eighteenth-century map of roads, where turnpikes netted the English countryside, into a nineteenth-century diagram of London linked by corridors to the capitals of its former colonies, Dublin and Edinburgh. In the management of these roads, a new form of government had been pioneered: centralized, expert driven, and capable of managing immense charges of information.

As a result of the tireless efforts of Sinclair and his Scottish allies in the Highland Society, Scotland was the first nation targeted by parliamentary committees. Through lobbying supported by the massive reports of the Highland Society, the Board of Agriculture, and parliamentary committees, Sinclair succeeded first in achieving a parliamentary board appointed to look into the problem of communications in Scotland and then in gaining control of that committee and its monies.

An 1803 commission began investigating both the original military roads through Scotland and the possibilities for new routes.[166] Telford found these roads crumbling and recommended immediate action. Made "at a time when Road making was little understood," the roads were now so steep by contemporary standards as to be "nearly impracticable, and so much exposed to injury from Mountain Torrents, that it may be demonstrably cheaper to alter [the road] than to continue to repair it Year after Year." Those roads followed routes impracticable for civil trade: "Military Roads may by possibility very properly pass along high ground, where for civil purposes a Road would never be thought of."[167] They also went through the wrong regions: constructed to police the glens of the east, they connected little, if at all, to the west coast, where development, ports, and new towns were looked for now.

Of all the centralized building of the nineteenth century, this first project was the most straightforward to fund. The army that had built roads into Scotland in 1750 in order to invade it was still paying money to keep up those roads in 1800. Arguing that its moneys could be better directed with a little organization was not a difficult step to make.

The commissions arranged for Parliament to fund the surveys and eventually the building of roads. Repairs and maintenance were originally supposed to be paid for out of the local purse.[168] Local authorities proved unable or unwilling to pay for the new level of maintenance required by government legislation. The local governments of Inverness, Ross, and Sutherland suggested in their letters to the committee that Parliament should pay for half the costs of the repair of roads, and Cathness "proposed to obtain Public aid exclusively" until its tolls should make a reasonable sum.[169] They were denied funding, and the committee warned them to repair their roads. This correspondence continued for a trial period, but, in the words of the 1811 committee, "The experiment has entirely failed, and the Roads first finished now evidently falling to decay."[170] New resolutions were drawn up, military moneys were rediverted, and the larger British public paid £3,500 of the yearly £8,000 spent on road maintenance in Scotland.[171] Road maintenance, as well as road building, had turned out to be such a massive expenditure that only the national government was capable of providing it, the 1813 committee concluded: "Such Management can only be established by a Legislative Interposition; which could not be successfully and satisfactorily exercised without affording some Proportion of Public aid, at least for a limited period of time."[172] Parliament had thus become the cobuilder and co-maintainer of 920 miles of road in the Highlands and a further 183 in the Lowlands.

By 1830, anyone riding north toward Scotland could travel on well-paved roads all the way to the Isle of Skye. Turnpikes under parliamentary control took the passenger as far north as Birmingham, or less far if he went the eastern route. But those turnpikes were slated for parliamentary control, and soon, some hoped, the same straight highways would shoot clear north to York and then to Edinburgh. In the Highlands, where the rocky hillsides housed only a few poor shepherds and crofters, travelers found signs of advanced industry. A total of 1,117 bridges connected straight, wide roads as far out as the islands, bringing them all into contact with the ports and capital, at a cost of £1,150,000, or roughly the equivalent of all tax revenue collected in Ireland in one year.[173] These parliamentary roads constituted about one-quarter of all roads in Scotland.[174] Wales and Ireland too were well connected. Iron and stone bridges with arches of spans as wide as 150 feet bridged the Dornoch Frith and the River Spey. These marvels of modern engineering contrasted with primeval territory "where the rapid river cleaves for itself a passage through the solid rock."[175] Parliamentary roads opened the Highlands to "shoals of travellers" every summer, who marveled at the remote vistas now made accessible to traders and tourists: "rugged heights, brawling torrents, and fearful mountain-passes, undreamt of out of the Highlands themselves half a century ago." Now these natural wonders were "exposed to the gaze of the wearers of pink parasols, and transferred to their scrap-books."[176]

The next stage of building connected London with the west. The committees of 1810–1811 and 1815 set out the case for building new roads in Wales and the north of England. In 1815, £20,000 was voted and an act was passed setting up the commissioners. In 1816, a further £10,000 was voted, and work began.

As in Scotland, the local authorities were supposed to take over road maintenance, but they turned out to be unable to manage it.[177] Individual trusts, whose stewardship of the public's investment in their roads was yearly reviewed by the committee and the chief surveyor, were found to be lax. Regular difficulties were exacerbated by the harsh winter of 1817, when corn riots appeared across the nation and poor Welsh counties were unwilling to provide any revenue for the maintenance of their roads.[178] The Holyhead Road committee channeled £6,000 to individual trusts in one year alone and then gave up. In the hands of local legislators who could not perform maintenance, the original investment of the build was spoiling. It was more cost-efficient simply to centralize the local trusts into one great authority.

Telford could point to narrow roads twelve foot wide that teetered over a hundred-foot chasm, supported by "very imperfect bulwarks, which are in danger of tumbling down." He could also, thanks to the standards for permanent road foundations that he and Macadam had been promulgating, characterize the entire highway as "so narrow, ill-constructed, and unprotected, as to be totally unfit for a great public Road." On the basis of Telford's testimony, backed up by elaborate reports gathered from his assistant engineers about every ten-mile stretch of the entire distance, the Committee on the Holyhead Road argued that the current highway existed in "so imperfect a state, that it must be entirely new made."[179]

The first stage of building in 1816 concentrated on five pieces of road held to be the most dangerous: the ascent at Bangor Ferry, the precipice from Ty-Gwynn to Lake Ogwen, the waterfall at the Lugwy, the precipice along the face of Dinas Hill, and the leveling of the steep path to Glen Conway. The projects involved major infrastructural configuration: high embankments, supported by strong masonry," "cutting through great masses of rock," and roads that clung "along the sides of rugged mountains." They were interventions of "considerable expense," but the committee had made a strong case for the vital importance of a road connection between London and Dublin and the impassability of the existing route. Once executed, the impressive projects in Wales justified requests for further expenditure, and the committee could point, in Thomas Telford, to a reliable individual whose engineering was the root of "the very skilful and superior" manner of building.

In 1819, Parliament subjugated the Welsh trusts under a single, centralized body of fifteen commissioners, administered by a Board of Commissioners appointed by Parliament, collecting sums for repairs out of tolls, and reporting back to Parliament on its progress.[180] The Board directed a chief surveyor and clerk who traveled the roads regularly, the surveyor looking for improvements to be made, the clerk collecting the tolls and examining records at each of the tollbooths.[181] The Welsh part of the Holyhead Road had been rebuilt and centralized under a model, expert-run authority.

Financed through these piecemeal grants, the Holyhead Road project grew in scope and solidity for two decades. In 1810, the committee began reviewing possible plans for bridges across the Conway and the Menai Straits that separated the Welsh mainland from the island of Anglesey, where the town of Holyhead held the best possible prospect for a harbor.[182]

Such ambitious plans and coordinated financing set in motion a rebuilding scheme of monumental proportions. Telford set to work filling in and leveling the rugged lowland hills where the road passed in Ogwen Bank, Owcu Glendwrs, Rhysgog, and Cernioge; rerouting the corridor around the "crooked and inconveniently narrow" streets of Llangollen Church; and embanking the deep valley between Chirk Castle Gardens and the village of Chirk.[183] The team embanked 1,144 yards of the Stanley Sands near Holyhead, where the sea came in over slippery slopes, and the sandy approach to the Conway Bridge, where the tide rose up to the level of the bridgehead.[184] At the "rapid tideway" of the Menai Straits Bridge, a "structure . . . of very unprecedented novelty and magnitude," consisting of two thousand tons of iron, arched over the waters there, soaring one-third of a mile from end to end. The results were impressive. As Samuel Smiles, still awed a generation later, put it, "Angles were cut off, the sides of hills were blasted away, and several heavy embankments run out across formidable arms of the sea."[185] Telford constructed forty miles of new road in northern Wales, aided by grants from the public purse in 1815 and 1819. Travel times were dramatically reduced between London and Dublin; by 1819 the Holyhead Mail performed the trip to Holyhead in thirty-six hours, "that is, in six hours less time than it did in 1815, and in 10 hours less time than it did in 1808."[186] In 1824, 1,754 carriages, chaises, and gigs were crossing through the tollgates of the Holyhead Road, and in 1828 that number had increased to 3,968.[187] Streamlined and popular, the Holyhead Road in Wales was a model of how parliamentary highways could serve the nation.

Successes in Wales precipitated action in England, where the other half of the Holyhead Road connected Coventry and Shrewsbury to London. Telford surveyed these hundred miles and reported to the commissioners on possible improvements in 1818.[188] Reviewing his evidence, the committee decided that the best course was, as in Wales, simply to take the road out of the hands of the local trusts. "Persevering in the old method," it wrote, was already resulting in "a great waste of money."[189] Consolidating the trusts under a new regime, parliamentary commissioners would reengineer their accounts while Telford reconstructed the physical fabric of their roads.

The proof that the commission provided better management than the trusts was plain. In 1821, the commissioners were managing a yearly £39,000 delivered from the individual trusts, in addition to another £8,298 in incidentals for which they applied to the Treasury.[190] By 1826, the commissioners could proudly report that they were taking in an annual £570,490 from the local roads and spending only £538,110, including a down payment on loans they had received from the Exchequer, with a profit of £42,380 for future improvement.[191] High returns from more successful trusts could be invested in improving the roads along the entire route.

Funded to make new improvements, Telford began overseeing the derelict roads around St. Albans, a diversion around the town of Wednesbury, and the leveling of the summit of Summerhouse Hill near Wolverhampton.[192] By 1830, even the English turnpikes of the Holyhead Road were administered by centralized experts and parliamentary fiscal oversight and were removed thoroughly from local control.

Meanwhile, the 1824 Holyhead Committee put some dozen turnpike trusts under the authority of parliamentary control. The commissioners brought the same predictability of numbers—of tolls, profits, expenditure on improvements, and standards for road dimensions, whatever their geography—to every trust they managed. Circular surveys were sent to all the trusts within the management of the Holyhead Road Commissioners to ensure better collection of data. These local trustees were polled: "What is the Amount of the principal mortgage Debt due by the Trust, and with Interest payable in respect thereof?" "What number of Loads, or Cubic Yards of Gravel or of Stones, have been put out upon the Road in the last year?" "Does the Surveyor attend to the rule for keeping the Roadhedges under five feet in height, and for removing trees?"[193] Elaborate tables displaying the interest, debt, and toll rates of each trust within their care formed supplements to all of the later Holyhead commissioners' reports. Armed with the knowledge of physical and financial variations, commissioners could begin to standardize the trusts and roads throughout their domain.

After the apparent successes of the Holyhead Road, new roads began branching north, cutting through English parishes and turnpike trusts to carve out new corridors to Edinburgh. From 1825, the Holyhead Road Committee investigated the possibilities for a separate road connecting Liverpool and London, extolling "the importance of this line of Road . . . on account of its communication with Ireland."[194] In 1826, Telford presented his survey of possible routes and estimated expenses, and in 1828, legislation was passed to build such a road.[195]

By 1834, the road to Holyhead through Wales, "the best in the kingdom," was a marvel of engineering when the parliamentary commissioners finished it.[196] Flat roads, "as even as a 'parquet,'" had been made by "cutting down large hills and filling up deep valleys" and ran past "floods of foam" in the "dark waters below," past "towering cliffs," "grotesque groups of upright rocks," and "somber woods, . . . here and there concealing the torrent."[197] The line included cuttings through sheer rock, "in some parts 30 feet in height, with high breast and retaining walls, stone parapets laid in lime-mortar. . . . So that this formerly frightful precipice is now a safe trotting road."[198] Tall embankments lifted the road above mountain cataracts and tempestuous coastal sands.

Unlike in Scotland, the finances for these improvements were entirely under government rather than local control. The government appointed certain tolls for each trust, which were used to defray a loan made by the Exchequer Bill Commissioners.[199] Massive grants in cash were paid for the major infrastructure projects along the way: £63,752 for the harbors at Holyhead in Wales and Howth in Ireland, £248,912 for the Menai Straits Bridge, and another £287,498 for other bridges along the routes.[200] Rough estimates of the modern equivalent of these sums would be in the neighborhood of £125,787,078, £491,120,487, and £567,253,318, respectively.[201] Some of the funds came from the tolls collected by the local trust from travelers who used the road; others were disbursed as grants-in-aid by the Treasury; and still others were paid against the loan by the Post Office.[202] The total amount expended on all bridges, embankments, and roads in both Wales and England came to £759,718, of which £338,518 was "granted by Parliament, without any condition for repayment."[203] The applications to the Treasury mounted in a predictable sweep: £8,298 in incidentals in 1821, £108,198 in 1832, and £27,871 in 1825.[204] These sums sat as a debt in the Treasury, adding to the hefty £5,028,129 debt that the Treasury had accumulated by 1827 from spending on public improvements.[205] Weighty public expenditures and massive public indebtedness characterized the age of improvement.

The triumphs of the select committees were the knot of roads that connected the Highlands and the direct corridor between London and Holyhead, each characterized by unprecedented scale, regularity, and expense. Those two projects, of course, were only part of the entire scheme of national reconstruction imagined by Henry Parnell, but they realized the grand project of colonial development advocated by John Sinclair and the Highland Society in the 1780s. Scotland, Wales, and Ireland were now concretely incorporated into the geography of British trade, mail, and defense.

The select committees also established general regulations for every turnpike and parish road in the nation. In the 1820s, major turnpikes began reporting their finances to Parliament and came under the control of engineers who were appointed by Parliament and reported to government committees.[206] In 1835, local authorities became responsible for following the "scientific standards" set by Parliament.[207] Macadam's "General Rules" and the precise equipment they demanded were mandated across the nation. Success for the highway lobby also came in the form of new regulations that tied private turnpikes into a closer relationship with government directives.

3

PAYING TO WALK

The National Movement against Centralized Roads

> The roads to [London] are level. They are smooth. The wretches can go to it from the 'Change without any danger to their worthless necks. And thus it is *"vastly improved, Ma'am!"*
>
> —WILLIAM COBBETT, *Rural Rides*, 1830

In 1843, Welsh coal miners began to sneak at night to the wooden toll bar at Yr Efail Wen in Carmathenshire that collected fees for users of the ancient road. The road was being paved for the first time, and in Wales, as elsewhere on both private turnpikes and Parliament-built roads, the cost of paving was defrayed by the tolls collected from the road's users. The Rebecca rioters, like participants in some dozen other riots during the past century, dressed up in women's clothing and left indignant biblical verses warning the toll collectors to stop their charges in the name of godly justice. Like the tarrings and featherings and "loud music" of early modern villages around Europe, these toll breakings warned the village rich to curtail their charges on the life of the poor.[1] But by 1843, a national discussion about the virtues of highways had complicated the debate. Depending on who the reader was, Rebecca might appear as the enemy of progress or liberty's defender against tyranny.

Many believed that new roads, whatever their expense, would enrich the poor. Radicals like William O'Brien had declared "liberty of locomotion" to be "one of the rights of man."[2] The sympathetic MPs who represented Scotland and Ireland agreed and demanded that Parliament subsidize the highways to their remote districts. To men like those who built the Holyhead Road, Rebecca represented a dangerous challenge to the state. The riots were a throwback, a sign that Britain had failed to understand the benefits of modernity and was taking up a form of government unlikely to ensure its continued rise. An anonymous civil engineer called Rebecca a "disgraceful" instance of "opposition to scientific engineering," the action of "ignorant obstructers," and the mirror of "barbarous ... legislation" that blocked the progress of industrial design.[3]

Charles Knight, the encyclopedist and historian, thought Rebecca no better than anarchy, feared the moment "when bodies of ignorant and lawless men feel their physical strength," and soon believed "that this power is sufficient to bid defiance to the law for the removal of every supposed grievance."[4] Most of the engineers and advocates of centralized highways whom we met in the last chapter would have struck a similar attitude toward the events in Wales. It seemed to many that more government control, more expert design, more policing, and more public spending would eventually relieve the country of such riots as ignorant peasants learned the value of progress.[5] The correspondent of the *Times* in Radnorshire impatiently wondered why the government did not merely take the turnpikes into its own hands immediately, thus demonstrating to the rioters that a gentler bureaucracy could work in their interest.[6]

Despite the promise of development, however, many viewed parliamentary highways and tollgates alike with suspicion. London reporters for the *Times* stressed the fact that the riots were a protest that intended to restore the ancient freedoms of the road; they emphasized the mob's descent on tollgate and toll bar and detailed the economic extortion of local people by a corrupt turnpike trust.[7] They argued that new roads—both turnpikes and parliamentary highways—unnecessarily increased expenditures, and that burden was carried by English taxpayers and the toll-paying users of the nation's roads. The riots against roads voiced legitimate concerns about expense. Welsh peasants needed neither turnpikes nor parliamentary revenue, but the right to determine who managed their local roads and how. Chasing the gatekeeper away and hurling pieces of mangled toll bar after him, this crowd offered a new icon of British modernity: the people were telling the bureaucrat to run.

In fact, far from provincial Radnorshire, other Britons were imagining Rebecca as the beginning of a national movement for popular control of the roads. In 1843, copycat crimes began appearing in North London, including the breaking of a new toll bar in the suburb of St. Cross, the property of a notoriously corrupt trust that had long been the subject of public complaint.[8] The neighbors left a piece of paper attributing the crime to

Rebecca. The *Times* applauded their prank and published evidence about the trust's mishandling of revenues: "If ever a public wrong, pertinaciously persisted in, justified a physical force remedy, the gate of St. Cross would be that wrong."[9] Throughout Britain, it seemed, local communities were demanding the right to determine the structure of roads for themselves.[10]

Through such statements, provincial coal miners in Wales were being transformed into an alternative icon of modernity. *Chambers's Edinburgh Journal* explained how "riot . . . procured that which peaceful remonstrance failed to accomplish."[11] The Chartist W. J. Linton, who edited the periodical the *English Republic,* called Rebecca "the one successful uprising in England since the Great Rebellion" and implied that tollgate breaking was one step on the way to universal suffrage.[12] Even Harriet Martineau, elsewhere impatient with rioting laborers who dreamed of bottomless wages, conceded that Rebecca was a "victory" of "humble farmers" over the "illegal" system of turnpikes that contrived to steal their hard-earned profits.[13] This side viewed concerns about the rights of those discontented with road management in Wales and St. Cross quite differently from the centralizers; rather than waiting for state intervention, they claimed, local people had the right to organize for reform. In other words, provincial issues of turnpike management had evolved into a debate of national proportions about the proper relationship between local organization and the state.

To understand the range of reactions to the Rebecca riots, we must begin with the deep history of the language of political participation and freeborn rights that sustained the claims of those who advocated local resistance. Eighteenth-century riots reflected the concerns of those least able to afford road tolls. The same concern for the poor users of roads reappeared in the 1820s and 1830s, voiced by radicals who feared that parliamentary committees were equally deaf to the concerns of the poor. Some, like the radical writer Richard Milnes, urged national ceilings on the tolls charged poor tradesmen for using the roads; others, like William Cobbett, pointed to the use of highways to benefit corporate or political interests rather than the declining agricultural interests more numerically representative of the people. Their warnings encouraged a climate of dissatisfaction with parliamentary road builders.

In the 1830s, politicians like Edward Knatchbull and writers like Joshua Toulmin Smith began to transform that discontent into a powerful new movement against the parliamentary centralization of roads and highways. Like the eighteenth-century villagers and radical writers who preceded them, these new localists grounded themselves in the participatory nature of local politics, which, they claimed, was more accountable to the communities it served than the "secret boards" of parliamentary committees. Unlike their predecessors, this movement was never intended to reform local communities. Rather, the newer voices took their most radical stand on the management of the state: whether rich parishes should be forced to subsidize the infrastructure of poor ones, as they had done for the Highland Roads and Bridges and the Holyhead Road in Wales. The new localists characterized the centralizers' practice of redistributing wealth as a source of inefficiency and corruption. They targeted the parliamentary, expert-led highway-building schemes of the 1810s and 1820s and argued for a devolution of political authority back to the level of the parish and county.

Readers of nineteenth-century history are familiar with parts of this story, namely, the movement of financial reform that targeted the military-fiscal state that followed Waterloo, and the growing conviction about the importance of local government after the public health reforms of 1848.[14] These episodes are typically treated as disjoint events, the creation of separate movements with little in common.

The story of road management, however, reveals a deeper continuity. Already in the 1820s, calls for financial reform targeted domestic road building at the centralized level. Fed by distaste for mounting parliamentary expense, a backlash against parliamentary road building was roaring by 1830 that resulted in the defeat of every single project for centrally managed roads that appeared between 1835 and 1900. In the 1840s and 1850s, the antigovernment message of the localists was carried into print, taking the form of a powerful, populist movement.

Local Politics and the Challenge of New Infrastructure, 1730–1835

The origins of discontent with the centralizers' system lay in a century of concern about road systems that depended on tolls for their revenue. In the century of turnpike expansion between 1730 and 1830, local administrators proposed a variety of regulatory responses, ranging from total collectivization to the centralized regulation of local turnpike finances. In reacting to turnpike rates and tollgate riots, local governments showed themselves eager to provide relief to the poor who were taxed for transport in the form of turnpike tolls and highway rates. The access to roads of the community's poor dominated early conversations about what kinds of roads Britain should have.

From the eighteenth century forward, tolls became the subject of changing forms of politics, including riots out of doors, local action, and centralized regulation, all of which aimed at the mutual end of freeing the ordinary traveler from prohibitive costs.[15] As tolls proliferated with the turnpike- and bridge-building booms of the 1740s and 1750s, travelers encountered them whenever they wished to cross a river or enter a town unless they were already citizens; the tolls were larger if a traveler had a horse or a wagon.[16] The London bridges charged a halfpenny even to foot passengers; only vagrants with passes and soldiers and electors going to the hustings glided through exempt from charge.[17] These tolls, visited on both the long-distance merchant and the short-distance commuter, were felt as a burden by the poor. Rather than pay, some travelers cut dangerous paths through sometimes-flooded sands along the coasts or broke through hedges.[18]

Local settlements had a variety of means at their disposal for protecting the interests of all citizens from the challenges of expanding communications. Local governments could preserve access to the roads by the poor through establishing local government and parliamentary authority over turnpikes. In Rye, for example, the corporation heard complaints in 1766 "that the tolls taken at the ferry at Rye Strand were very exorbitant" and threatened to petition Parliament to revoke the toll collector's license unless he lowered his rates.[19] Petition and local management formed the lightest echelon of government intervention.

Local governments could extend those freedoms even further by replacing a private toll road with a public one. Local governments often preserved access by buying out the interests of private communications developers. In 1723, a new bridge in Rye passed the Edward's Dock and put out of business Mr. Robert Hounsell, the owner of a ferry that had operated for 150 years. The commissioners of the harbor compensated him £10 and his tenant £8 against an annual rent of £40. In 1725, a new road built by the corporation allowed the inhabitants of Rye to bypass the private ferry, and the commissioners of Rye Harbour paid £4 compensation to Lady Doneraile, the ferry's owner.[20] The Isle of Man subsidized its roads with a tax on hunting dogs and publicans.[21] In Buteshire, tolls were avoided through the private largesse of the Marquis of Bute.[22] By buying out private interests, local governments directly intervened to relieve the poorest citizens of tolls and taxes without jeopardizing the development of infrastructure for all.

Where local municipalities failed to intervene, "active rebellion" took the form of turnpike smashing. The incidents peaked in Gloucestershire and Herefordshire, where poor farmers were unable to pay the tolls collected from wealthier traders passing between markets. In 1735, the "Ledbury Rioters," "a hundred in a gang, armed with guns and swords, as well as axes to hew down the turnpikes," filled the county jails, and their leader was hanged at Tyburn.[23] Large riots appeared in 1749 Somersetshire and Gloucestershire, where the rioters blew up the toll bars with gunpowder.[24] According to William Albert, the historian of these riots, the resistance aimed to sustain cheap access to roads that had in recent memory been "free to all." The riots enjoyed "passive support from the community at large" and succeeded, Albert suggests, in awakening a sense of moral responsibility among local magistrates and trustees with regard to exorbitant rates.[25] Such regular outbursts in the eighteenth century continued into the first half of the nineteenth, the work of locals endeavoring to ensure that the poor continued to have cheap access to the roads immediately around their homes.[26]

Patterns of protest and reform continually negotiated rights of use and access at the local level. This moral consensus about the need to adapt infrastructure to the principle of access existed without direct parliamentary intervention and indeed in the absence of a coherent political discourse about the nature of the interventions. A theory to unify, explain, and extend the principle of access would be provided only in the 1810s and 1820s, as parliamentary activity in road building made the whole-scale management of roads a matter of national concern.

Tolls struck some contemporaries as a limitation on their freedoms, a challenge to trade, and a burden on the poorest members of the community. As centralized government began to reconfigure the management of roads after 1796, Parliament replaced local legislatures as the arbitrating authority over questions of access. While Parliament began turning both turnpikes and parish roads over to a centralized bureaucracy, radical writers like William Cobbett, Richard Milnes, and William O'Brien translated the concerns of eighteenth-century communities into print as a call for extending access. Within the radical press, the calls for access took on a new specificity and materiality that made explicit a critique of expert rule.

Radicals frequently complained about the arbitrary and frequent provision of tollgates in the absence of government regulation. Milnes explained, "I occasionally visit a friend at the distance of five miles, and have to pass three bars, side bars, or chains, on three roads, and they take from me 4¼ d., though I do not travel much above half a mile upon all the three roads."[27] *Chambers's Edinburgh Journal* described "a mesh of thirty-five toll-bars and checks" around Edinburgh, including "several indecently placed within the streets of the town."[28] Arbitrary injustice and frequent tolls made the turnpike system an obstacle to the fair distribution of the burden of the nation's improving roads.

Radicals gradually developed this critique of uneven access into a picture of how turnpike tolls served as an unjustifiably burdensome tax on the poor. Cobbett complained that "the turnpike toll for the poor man's *ass* is the *same* as for the *hunter* or the *racer*, or *carriage* horse of the lord."[29] Milnes was outraged to see a "poor chimney sweeper," who had not enough money to buy "halter or shoes," charged twopence at the turnpike at Westgate Bar, near Wakefield, the same rate that Milnes himself paid for his horse.[30] Cobbett suggested the reform of the "toll on Asses" because "the owners of these animals are generally very poor people, who have not the ability of purchasing or maintaining horses." He reminded the trustees of "the trifling injury done to the roads by their light pressure, and the slow rate they travel," suggesting the injustice done when the weight of financing roads fell hardest on those least able to afford it.[31]

Radicals extrapolated from a logic of just taxation to the existence of universal rights of access to the ancient routes of the kingdom, which had been open before they were improved. William O'Brien explained that a turnpike gate "operates not only as a tax upon those who use a road, but also as an impediment which deters a large portion of the community from enjoying pleasure and advantage which would otherwise be open to them." "A turnpike gate," he declared, "is an odious interference with the liberty of locomotion, which is one of the natural rights of man."[32]

By the early nineteenth century, radical calls for the gradual abolition of road tolls were being recited by engineers and highway advocates as among the possibilities a future-looking scheme of road design could offer. Thomas Estcourt, writing in the 1790s, favored a reform movement that would rationalize taxes across the nation to fall most heavily on those with the vehicles that harmed the roads the most.[33] Road reform was in the air, and a general climate of expectation was rising.

When parliamentary road builders found it necessary to subsidize their expensive projects with tolls, they disappointed such hopes, creating a climate of dissatisfaction with centralized government that would fuel other critics. Centralized road building, like that executed in Scotland, Ireland, and Wales between 1796 and 1835, tended to favor colonial peripheries where new roads could aid the cause of national assimilation.[34] These plans tended to favor joint public-private road improvement in the service of rapid infrastructure development, and thus subsidized turnpikes under state control typified both Scottish roads and the Holyhead Road to Ireland. By 1835, forty years of centralized road building had failed to provide open access to the nation's highways.

But ambitious plans pointed to an era when centralized taxation and administration would ensure access and improvement across the entire nation. Henry Parnell's plans for a centralized board of roads were mooted in the 1819 committee on turnpikes and highways and detailed in his 1833 *Treatise on Roads*.[35] Others followed, criticizing turnpikes and local governments as hopelessly entangled relics in need of external reform. They frequently provided detailed, quantified reports in which they argued that tollgates sucked up more money than they spent on road repair. William Pagan explained in a treatise on reform that in Fife and Kinross, less than half the tolls collected on roads were actually expended on road repair, not including the expenses of introducing road legislation in Parliament.[36] Pagan thought that this system of local management gave rise to local enmity and further retributions, with smallholders evading tolls and toll collectors attempting to take more tolls than were due to make up the difference, and both parties frequently ending up in court. A journalist for *Chambers's Edinburgh Journal*, endorsing Pagan's books, argued that turnpikes resulted in a "world of small litigation" and explained that "the whole road system of Great Britain, with its eight or nine thousand managements, its endless exactions, and its universal network of toll-bars, is, without exception, the most awkward and absurd institution on the face of the earth."[37] For centralizers, the abolition of tolls represented an opportunity for the rational exercise of power from above typical of the best uses of government bureaucracy.

The bureaucracy itself raised the expectation that if it were given power, it would do away with tolls. Representatives from the Post Office, interviewed in the highway committees, testified that by their calculations, local turnpike administration had proved itself inefficient and wasteful. "At an annual expence of two millions for repairs," they still could not secure "the safety of the traveling part of the community . . . constantly exposed to dangerous accidents," and "conveyance" was still "performed at a very heavy expence."[38] John Holt, the Board of Agriculture's correspondent for Lancashire, pointed out that turnpikes had already raised "large sums . . . to render the general communication easy and certain," and he advised that the public was the best possible administrator of a project of such scale.[39] William O'Brien also thought that rationalization would help the turnpikes. "A local rate of £300 could in most cases be collected at an expense not exceeding, at the utmost, one shilling in the pound (or £15;) whereas, if that sum be collected from the public by means of a turnpike toll, it is probable that, at least £1000 will be absorbed in the mere expense of collection."[40] Radicals and government officials alike agreed that turnpikes represented a challenge to the public interest.

Concerns about the poor paying for roads, first articulated by radicals, were soon echoed by local surveyors and parliamentary representatives. Richard Bayldon, a road surveyor on the Sheffield and Wakefield Turnpike, writing to his trustees, encouraged them to readjust their tolls. He framed the conversation within the renewal of the Turnpike Act, an occasion, he explained, when it became "desirable to deal justly with all parties as respect the tolls."[41] Officials frequently understood turnpike reform from the rationale of community wealth based on free circulation. Echoing Adam Smith, the 1836 Select Committee on Tolls and Turnpikes pointed to the importance of a nationally integrated market to the wealth of the entire nation: "In a civilized community, . . . every individual must more or less be benefited by the facility of communication," and tolls represented an "impediment" to the "free intercourse of commodities."[42]

Generalizing from traditional values, later writers extrapolated the existence of a freely circulating community whose informal travel and trade were threatened by the mismanagement of roads. The *Times* diagnosed the psychology of the toll collector as a soured, bitter, vengeful failure trying to revenge himself on the character of the natural Briton, who hated nothing more "than to be stopped in his progress."[43] Lobbying for toll reform in 1860, Mr. Alcock denounced his opponents as defenders of the "capricious exercise of power."[44]

Concern with the adverse effects of tolls on the poor was generating a search for new models of road revenue within Parliament. So solid was the conviction about the importance of access that by the 1830s it had spread to infect even the highest echelons of the civil service. Colonel J. F. Burgoyne, chairman of the Board of Public Works in Ireland, testified before a select committee in 1836 that turnpikes created "a perpetual impediment and interruption to the progress of business and recreation," and that "however small" the toll, it was taken in an "inconvenient manner."[45] Even James McAdam, who carried on his father's work as an engineer for turnpike trusts, likewise agreed that tolls were "an inconvenient mode" of raising revenue.[46] When the issue was raised again, the 1840 Committee on Local Taxation undertook the first total census of the nation's roads, aggregating the length of public and private roads, the number of gates, the condition of the roads, and details of financing and management for every road in the country. They found 1,116 trusts governing 22,000 miles, with about 7,796 tollgates, governed by 3,555 treasurers, clerks, and surveyors. Parish highways, by contrast, governed 104,770 miles under 20,000 parish surveyors or way wardens.[47] Such sums of money, officials reasoned, could be better invested in improved roads than in the provision of additional gates and payment of tollgate keepers.

Given the state's position, it was reasonable to expect that Parliament would abolish tolls on the Holyhead Road and the Highland Roads and Bridges. But mounting costs, driven by the expensive embanking of routes and raising of monumental bridges, drove improvement expenses ever higher. Parliamentary commissioners kept the tolls in effect as a means of reducing the cost. In the 1820s and 1830s, travelers were still encountering tollgates on the national corridors; the promises and hopes of radicals, echoed by engineers and bureaucrats, were being failed by the mechanisms of the state.

In this climate of dissatisfaction, experiments with raising revenue by other means than tolls were appearing in localities around the nation. In London, for instance, the Metropolitan Road Trustees charged the asses ridden by the poor "one-half toll of horses," a provocative gesture of inclusion, described by radical Richard Bayldon as "an act of common justice, worthy of imitation."[48] Many of these attempts echoed the centralized model in consolidating tollgates so as to eliminate the high cost of toll collection.

Lord Lowther encouraged the city to consider consolidating its tollgates into a ring of bars around the city that would remove the charge from the urban dwellers and place it on the shoulders of outsiders.[49] A parliamentary commission recommended an act of Parliament to consolidate all the trusts in the county of Middlesex under a single private trust, administered by local individuals rather than parliamentary appointees. London parishes abolished their tollgates, one at a time, between 1822 and 1873, as individual localities voted to join the Metropolis Turnpike Trust.[50] The London commission's activities, a departure from the centralized, Parliament-run toll roads of the 1810s and 1820s, stood out as an example of how local entities could pioneer road reform without the national government. The Metropolis Turnpike Trust enacted road reform through government, but at a local level. Its activities expressed an avenue for public dissatisfaction with the way tolls and charges had been pursued by the Commissioners of the Highland Roads and Holyhead Roads.

The London pattern of voluntary, local abolition spread through various counties on a piecemeal basis. These events set a compelling model for other local governments attempting to displace the burden of road payment.[51] In Montgomeryshire in the 1830s, locals petitioned Parliament for an act to consolidate the "whole of the roads," breaking up the local turnpikes into four highway districts "according to the convenience and locality."[52] Local authorities, stunned by the persistence of tolls on parliamentary roads, were sure that they could charge their users less.

The Emerging Critique of Centralization, 1816–1848

It was in this climate of experiment and reform, with widespread anxiety about centralized bureaucracy's relationship to tolls, that a new movement emerged. Writers and politicians opposed to centralized roads began to accuse centralized government of having put into place an inefficient system. Centralized government, with its massive chain of command, they argued, propagated interestedness and corruption.

The first activists of localism were MPs suspicious of the way parliamentary funding was being reconfigured in the age of reform. Men like Edward Knatchbull and Lord Lowther, Englishmen from rural provinces who cultivated ties to the Tory Party, became vocal critics of centralized administration.[53] Lowther even engineered new projects for road management at an exclusively local level, such as the Metropolitan Turnpike Trust in London. Their campaign departed from a wider trend of challenges that questioned state spending, the so-called reform movement of the 1810s and 1820s by which Whig, Tory, and radical politicians attacked the excesses of British military spending.[54] In the hands of Tory politicians, however, the critique of spending took a new form. It was focused first against infrastructure and second against centralization in general.

Gradually, Tory politicians were joined by writers and journalists who actively articulated the grounds for local rule and developed an all-out critique of centralized government and expert knowledge. Prolific authors like Joshua Toulmin Smith and Elijah Hurlbut grounded localist challenges within a history of long-term Anglo-Saxon freedom, and targeted efficiency and accountability as faults implicit in centralized rule.[55] By 1835, a full-blown attack on centralized administration was in place.

The assault on centralized infrastructure began in the 1810s with the seemingly limitless bills attached to large infrastructure projects. In 1816, Knatchbull attacked the Ramsgate Harbour bill and stopped it from passing.[56] Lord Lowther challenged the continued sums diverted to the Menai Straits Bridge on similar grounds: mounting costs consistently overturned original expectations about the nation's financial responsibility for these projects. In 1830, challenging another request for cash, he pointed out that the Caledonian Canal, projected at £474,000, had cost the nation £961,000 by its completion; the Highland Roads, projected at £96,000, actually cost £240,000, in addition to another £5,000 yearly for maintenance. The Menai Bridge, projected at £60,000 in 1815, had required £247,000 by 1823. Lowther voiced the obvious: he "wished to caution the House against those Parliamentary Commissions, and not to place any faith in the prospects they held out."[57]

By the 1830s, these charges of inefficiency were so prevalent that they were being voiced by crucial witnesses in the select committees themselves. Colonel J. F. Burgoyne of the Board of Public Works in Ireland, the first witness called for the 1836 turnpikes committee, argued that the centralized administration of roads would deprive the "locally concerned" "public" of its entitlement to participatory politics in the form of "direct supervision and control over local expenditure."[58]

Localist writers frequently drew attention to the power of centralized committees to reroute commerce in their own favor. *Blackwood's Edinburgh Magazine* pointed out how "the Chairman of the Morpeth and Wooler Road Committee, who is resident in and intimately connected with East Lothian, declined to have anything to do with the Committee, until he should be satisfied that the improvement of the Wooler road would not be materially detrimental to the interests of East Lothian." As a direct result of his intervention, the committee rerouted the road between Morpeth and Edinburgh so as to ensure the "profit" of the towns of Alnwick, Berwick, and East Lothian. The magazine encouraged its readers to inform themselves about all centralized pleas for road inform, to look at the surveys for themselves, and to remain vigilant.[59] In the 1830s, Lord Lowther challenged the Holyhead Road Committee to justify the expense of northern roads from the point of view of a larger nation. The funding passed, but by a margin of only thirty-four to twenty-seven, suggesting that localists spoke in terms that were convincing to a significant proportion of the members.[60] Localists questioned the logic of "improvement" and drew attention to how pleas for centralized management of infrastructure opened opportunities for experts and advocates of centralization to enrich themselves from the public purse.

In contrast to the dangers of centrally planned roads, localists pointed to the instinct for piecemeal economic improvement that dominated in local government. In the committees of the 1830s, numerous surveyors testified about the interest of local landlords in improvement and the lengths to which these parties would go to help develop infrastructure when they were left free from outside interference. Thomas Penson, county surveyor of the Denbighshire roads, testified in 1836 that the 130 miles of roads being developed under his care went across land for which he had paid almost nothing, even when the road cut through "lands of very considerable value," because the "proprietors and persons in the neighborhood" took a "great interest" in the improvements. He argued that "the gentry of the neighborhood" were the chief reason that he had had such liberty to pursue the local improvements at this scale, and that a centralized board of management running the Denbighshire roads from London would "negative the feeling for improvement which at present exists on the part of the gentry."[61] Even James McAdam, whose father had made his fortune working for Parliament, took the line that local government would be more efficient in the future, arguing that roads "can never be placed anywhere with so much advantage" as in "the hands of the country gentlemen and local trustees."[62] The self-interest of local landlords suggested that they were independently motivated to ensure local improvement without outside intervention.

Government infrastructure spending had a tendency to skyrocket once a project had been approved. Rampantly spiraling inflation was not an uncommon occurrence in government accounts. The Ramsgate Harbour, for instance, came up for review in 1817. By 1816, the engineers had spent £370,000 on a project that John Smeaton had originally estimated at £17,000. Of the £10,000 appropriated in 1808 to build a harbor in Holyhead, not a penny went to harbor building; instead, the money was siphoned off to other projects.[63] Government infrastructure budgets had a way of inflating without oversight to the enrichment of private engineers.

An even more outstanding case of jobbery was that of Colonel Upton, a surveyor for the Holyhead Road Commission between 1818 and 1826, who was discovered in April 1826 to have siphoned £2,000 of public funds into his own purse. Charged by the assizes with fraud and forgery, a hanging offense, he somehow escaped in the night to Russia, where he became the emperor's engineer for the ports in Sebastopol.[64] Celebrated cases of unregulated expenditure damaged the image of the expert and his reputation as a disinterested servant of the public.

Beginning in the 1790s, localists opposed these propositions on the grounds of their being "jobs"—a term that referenced the bipartisan reform critique of ancien régime redundancies while simultaneously pointing to the fact that Scottish MPs seemed continuously to give national engineering jobs to Scottish engineers. In 1825, when John Loudon Macadam's select committee moved for the abolition of tolls in the Metropolis Turnpike Trust, Knatchbull accused the expert-run, parliamentary project of corruption and jobbery, "which he had always opposed."[65] Centralized contracts, localists contended, were as liable to corruption as cheaper local ones and were even more dangerous.

Many localists simply questioned the need for expert management of the roads at all. Some challenged the centralizers' assertion of expert impartiality. Lowther discounted the evidence put forward by highway experts, arguing that the committee "had referred to numerous memorials, but they were all founded on interested motives."[66]

Even more skeptical writers conjectured that the "general rule" propounded by experts was itself problematic. The concept of a general rule for infrastructure across a diverse nation seemed counterintuitive to many. "One District has a wet climate; another a dry one," admitted a writer for the 1811 committee. "One District has a light soil; another a deep one; one District is hilly and mountainous, another flat and low; one District has good materials for making Roads; another may be deficient in these important Articles; one District is populous, and has a great concourse upon its roads; another is but thinly inhabited, and has but few travelers."[67]

When skeptics looked at the carefully designed examples of parliamentary roads around them, they were frankly unimpressed. Testifying for the Select Committee on Tolls and Turnpikes in 1836, Colonel Burgoyne refused to admit that expert-driven engineering necessarily produced more perfect roads: "The macadamized roads have only been introduced of late years," and engineers everywhere were "far from understanding thoroughly the proper principles of their construction."[68] State experts were blind to local peculiarities, they argued. The *Edinburgh Review* conjectured in 1864 that "Central Government, apart from other almost insuperable objections, can never possess such a knowledge of local circumstances as would enable it to act satisfactorily, except by way of general regulation or supervision."[69] Joshua Toulmin Smith pleaded for his readers to see through the justification of expertise by those who calculated efficiency from the "mere material point of view" that reckoned "excellence by cheapness only."[70] Localists questioned the status of expert knowledge and argued for local experience as an alternative.

Critics of centralization pointed to the fact that basic infrastructure, theoretically of benefit to the national wealth of all, was of limited immediate benefit to the poorest members of the community. Radical journalist William Cobbett concurred, calling the highway projects "a *waste*" and "means *misapplied.*"[71] Cobbett described this process as an economic exchange with abominable consequences. While the provinces starved, London flourished. "The wretches can go to it from the 'Change without any danger to their worthless necks," he complained.[72] Richard Milnes, a trustee of the Barnsley and Grange-Moor Turnpike, explained that from the point of view of the laborer walking to work, "a winding path . . . a little rising and falling" was even better than a straight one, for it broke up his journey by "changing his position."[73] True improvements, radicals suggested, had to be analyzed from the viewpoint of their benefits to all classes.

Borrowing from this radical critique, localist writers questioned the need for indiscriminate development from above. Cobbett observed that infrastructure tended to benefit the wealthy alone, without regard for the poor displaced by those energies. The Edinburgh *Lounger* caricatured the centralizers through the persona of Mr. Dormer, the ambitious improver. "When any plan is proposed, which by theoretical deduction it can be shown may possibly be attended with some general advantage, but which will certainly be hurtful to some individuals, Dormer is sure to give it his warmest approbation and support." Sure that he is acting in the public interest, Dormer "thinks nothing of demolishing houses, rooting out inclosures or dispossessing tenants." The article continued, "I have known him, for the purpose of widening a highway only a few feet, pull down a house by which a widow and a numerous family of children were turned out to the open air."[74] By throwing the status of improvements in infrastructure into jeopardy, localists could effectively question the importance of expert rule.

Other voices protested that expenses tended to grow as government burdened one set of taxpayers with the debts accrued by another set. James Brook, a commissioner of roads for the Yorkshire and Lancashire roads, testified to local "alarm" at the parliamentary mooting of propositions for a consolidation trust bill that would forcibly consolidate the well-managed roads of Leeds and Otley with other trusts that had been poorly managed and so had been encumbered with debt.[75] Officials voiced fears that excessive private building would drain the public purse were the public to take over private infrastructure wholesale. James McAdam testified that much of the turnpike debt in the nation derived from intensely local road building "from gentlemen being desirous of making roads for the improvement of their estates," such that extremely expensive roads to serve these individual interests had been constructed without expectation of the bonds for the roads "ever to be repaid." He reckoned that plans for Parliament to repay all such debts on all such toll roads out of the public purse would be "unfair" to the "public," who had no reason to become "parties to the payments of such debts."[76]

Localists favored local knowledge and government over expert rule administered at a central level. Fears about the spiraling cost of pet projects were directly stirred by parliamentary building in Scotland, which struck many as an outright example of regional pork barreling at public expense. In 1830, Lord Lowther voiced his outrage: "There was no concealing the fact that this was a downright Scotch job, to enable Scotchmen to mend their own roads with English tools, that they might walk up from Edinburgh to London by a shorter way than at present." George Lamb, from Westminster, affirmed that it appeared to him "a Northern job," and that the petitions for the route had all come from Edinburgh.[77]

Lowther represented a rising tide of public opinion that closed these questions firmly with a blanket answer usually phrased this way: "He that uses the roads should pay for them." Local authorities should pay for and administer their own roads. This principle would be quoted in parliamentary committees as a tenet of political right for the rest of the century. It directly countered the centralizers' hopes of promoting collective responsibility for the roads. "Having tasted of the benefits of improvement, the knowing Southerners wished to monopolize it to themselves!" exclaimed the historian Robert Kemp Philp.[78]

Localizers countered such claims about collective responsibility with a theory of government accountability and the perils of large-scale management. Joshua Toulmin Smith conceived of centralization as a political poison for which local government was the antidote. He outlined how centralization worked by oligarchy, deceiving the public, and "secret Boards." According to Toulmin Smith, centralization was orchestrated by power-hungry bureaucrats who organized public "panic" and then moved in to enforce order; this was the case, for example, with Edwin Chadwick and his campaign for public health.[79] It operated by the permanence, expense, and scale of its undertakings. "Make your encroachments boldly," he wrote. "These encroachments . . . will grow into precedents, every one of which strengthens your power, and causes Free Institutions to become more and more but a name."[80] Toulmin Smith accused parliamentary advocates of centralization of meddling and ignorance: "The Members who complain of the state of the Highways, ought to know that, if they fulfilled their own duties in their own neighborhoods, instead of uttering complaints in the House of Commons, the means already actually exist to bring about every wholesome remedy."[81] Centralizers had succeeded thus far, Toulmin Smith warned, by promoting a dangerous lie of the necessity of expert building to economic well-being.

In contrast with the secret boards of centralized government, local government was presented as the epitome of openness. Parish government, Toulmin Smith wrote, "controls" its officers and remains "connected with the general well-being" of the population.[82] Referencing local road making in particular, he explained the relationship between community and management: "Whoever fills such an office should be able to count upon the cordial aid of his neighbors."[83] In local government, the major qualification for these offices was sociability and responsibility. For Toulmin Smith, a good leader was not an expert skilled in maths, but a man who could "count upon the cordial aid of his neighbors."[84] Such an emphasis on consensus was totally missing in the centralized administration of roads.

Others argued that local government produced accountability. W. J. Linton, a Chartist and the editor of the *English Republic,* argued for a system where local officers collected taxes and used them to build roads. "All these things," he wrote, "are strictly within the province of the local Government, and concerning them the State has no jurisdiction save as a court of appeal, so long as they do not counteract the general scheme and rule of the whole nation."[85] William O'Brien explained that local volunteerism, instead of a civil service, represented "one of the noblest characteristics of Great Britain," evidence of "that manly and self-relying spirit which is generated by the exercise of social duties of a public nature, and which is the best guarantee for the preservation of a nature's liberty." From this understanding, O'Brien reasoned that "turnpike tolls ought to be avoided as much as possible, and the public roads of a district should be provided for by means of local rates applied under the superintendence of 'district councils' representing the rate payers."[86] "What is needed in all these things," Toulmin Smith explained, was a fuller sense of "mutual responsibility, equally of those concerned and of those entrusted," and a clearing out of "irresponsible parties . . . to interfere in any way."[87] Free roads, Elisha Hurlbut explained, meant free men, and free men naturally gravitated toward small, independent communities in which their voices could be heard.[88]

Localists accused centralizers of holding themselves aloof from the simple responsibilities of accountability. The Edinburgh *Lounger* condemned the improvers who "arrogate to themselves the praise of public spirit, and look down with contempt on the humbler virtue of such as are occupied in the private concerns of life," concluding that "they are not quite so remote from selfishness as they would sometimes have the world to believe."[89] Joshua Toulmin Smith, the defender of the parish, even went so far as to link every badness in the roads to dependence on centralized government, which resulted at home in "the weakening of the full sense of responsibility" among "officers and vestries." He defined such expert-driven rationalization of the roads as prioritizing "selfish material concerns as the sole aim and end of life," a reflection, he argued, of their own debased morality.[90] Localists argued that centralization operated by deracinating knowledge from the society it was supposed to serve and cheating representational government of its intended effects.

Finally, localists generalized these tenets of openness and accountability as the basis for their understanding that local government best protected the ancient rights of Britons. Elisha Hurlbut, in his *Essays on Human Rights,* argued that county government of the roads would at once "*de-centralise* government, and give to individual man first, and next to small communities, importance and independence."[91] In contrast to centralized opportunities for authoritarianism, localists depicted parish government as characterized by familiarity, openness, and oversight. Under local government, wrote W. J. Linton, "There will no longer be any purchasing of freedom."[92]

The Triumph of Localism, 1835–1848

The result of this swelling tide of local engagement critical of centralized government was transition from centralized to local rule midway through the century. Following an alternative pattern of turnpike abolition, local governments fashioned a piecemeal approach to their local roads and gradually whittled away at the power of the turnpikes. Early successes around London and Scotland encouraged contemporary enthusiasm for the localist movement.

As early as the 1810s and 1820s, even as highway centralizers diverted millions of pounds to the construction of parliamentary roads, localists were demanding high levels of accountability and revision. The 1810 highway-centralization reforms had to pass through seven revisions before, in attenuated form, they resulted in some insignificant measures governing the shape of wheels and requiring turnpikes to provide footpaths alongside the carriageway.[93] After the development of a political critique of centralized interestedness and inefficiency in the 1830s, the era of localism reached its triumph.

By 1835, suspicion of highway centralizers, once voiced only in parliamentary committees, reached a new pitch and came to characterize the official statements of the select committees themselves. Thomas Grahame, an advocate of centralization, was shocked by the "*great distrust*, on the part of the community, in the government; and in its justice, and proper management."[94] In 1835, Mr. Portman, the MP for Dorset, found so much resistance to new highway legislation that the only measures he could introduce were permissive clauses intended to curtail the most blatant abuse of property, the stopping up of public highways by magistrates.[95] In the epoch that followed, not a single centralizing measure passed. Under the reign of the new localism, parliamentary legislation was characterized by the turnpike committees of 1836 and 1864 that rearticulated the importance of local administration, as well as the vigorous thrashing of new plans for centralized rule whenever they were mooted in public. The triumph of local government effectively ensured that one region was not asked to foot the bill for another region, and it limited the powers placed in the hands of anonymous parliamentary experts.

The new generation of centralizers abandoned ambitious plans for the expansion of infrastructure and grounded their activism in far more modest calls for the centrally directed abolition of tolls. They pointed to central administration as the most rational and effective approach to abolishing tolls and urged the prompt assignment of turnpikes to a parliamentary committee equipped with powers to arrange for their expiration. In 1845, William Pagan of Curriestanes, Kirkcudbrightshire, a banker in Cupar-Fife, presented a proposal for employing the Post Office as a central taxation authority to fund local improvements and the abolition of tolls throughout the nation.[96] Pagan proposed that a public tax, rather than tolls on all of the road's private users, finance the maintenance of the road network. The public tax would be applied to horses and thus fall most heavily, Pagan suggested, on the individuals who were most able to pay, rather than on the free laborer walking from place to place.

Brief waves of popular enthusiasm inspired briefer waves of legislative action. Alexander Mackinnon, MP for Lymington, was responsible for several bills in the 1830s to consolidate turnpikes into a single board managed in London.[97] As late as June 1835, centralizers were still trying to use the General Turnpike Act of 1773 to enforce the preeminence of a single centralized board of highways with control over all turnpikes throughout the country.[98] After Pagan's book went into its third edition, Lord Elcho, MP for Gloucestershire East, drew up a bill for the abolition of general tolls that was briefly greeted with enthusiasm.[99] The *Journal of Agriculture* reported that "public interest" was "thoroughly awakened" and predicted that "there must be a change." The costs were expressed in rising fees for short-distance travel that small farmers and tenants conducting their own trade found hardest to bear. A royal commission was appointed to investigate. Those who favored the bill touted it as a direct expression of "the feelings of the public."[100]

These advocates of centralized highway management did their best to link the cause of centralized toll abolition to popular expression and free trade. The advocates, typically Scots members of Parliament or lifetime civil servants, proposed models ranging from the national collectivization of roads beneath a single board to county schemes that would help local parishes collaborate in tollgate consolidation. They regularly employed arguments about collective utility that clashed with the localist instinct for self-determination. In 1850, George Cornewall Lewis, a lifetime civil servant and the author of official reports on the administration of Ireland and Malta, as well as several historical treatises on government administration, drew up extensive plans for a nationalized board of roads that would put all the nation's highways under control of Parliament.[101] Extrapolating from Adam Smith, Lewis argued that "the improvement of roads and bridges . . . annually increases the intercourse between citizens and renders their relations less dependent upon the arbitrary boundaries."[102]

Indeed, centralizers drew tight connections between free trade and centralized infrastructure as reform efforts that promised to increase the wealth of all parties. This link between free roads and free trade originated only in the 1820s, as the efforts to centralize roads were taking off.[103] In 1847, Lord Palmerston explicitly linked the notion of free trade with turnpike abolition and involved himself on those grounds in helping the cause of the Metropolitan Turnpike Trust.[104] In 1849, Robert Peel even proposed centrally directed turnpike abolition as the best possible compensation to the farming interests who would be expected to suffer the greatest setbacks as a result of switching to a free-trade policy.[105] Trade periodicals like the *Draper and Clothier* proclaimed that "the abolition of all tolls or turnpikes" was the direct outcome they prayed for from the free-trade interest in Parliament.[106]

Centralizers insisted that the parish system tended to engender abuses of the highway rates. They drew attention to the complicated system of gathering and assorting the rates, where poor rates and public health taxes were often put into the same pool. The *Edinburgh Review* complained, "We have heard of a church rate being paid, of overseers and vestry clerks receiving salaries, of a poor-law guardian being provided, even of a bastard child being maintained, out of the highway rates; the latter for the appropriate and cogent reason that it was alleged to have been found in a ditch by the roadside."[107]

Finally, advocates of centralized reforms downplayed the extent to which their plans would take authority away from the parish. In describing one such centralizing bill, Sir George Grey, its advocate, urged that "parochial rights were not confiscated"; "the money raised within a parish would be spent within that parish," and the district board would be stripped to its minimum.[108] Recommending some districtwide redistribution of highway funds, the 1840 committee took pains to prove that small parishes would benefit from these aims.[109] The committee insisted that its legislation would return power to the local level and hold the aggregated districts responsible to each vestry and parish of their subdivisions.[110] After 1850, every centralizing proposal offered only an attenuated version of reform meant to consolidate parishes at merely a county basis. Even these semicentralized reforms, however, would be abused and defeated.

Opposing these proposals were localists who viewed every form of centralization as an encroachment on the ancient rights of the parish.[111] The localists leaped on centralizers' attempts to legislate further controls over roads. The original wording of the 1835 bill, for instance, had given the commissioners of turnpikes power "to suspend at any time the execution of all orders to the present Trustees" and thus the power of veto over individual turnpikes.[112] Clause 34 proposed that all new turnpikes first be consented to by the commissioners. An anonymous indictment of the centralized powers so given called them "an unconstitutional denial of the rights of the subject, hitherto unknown in Parliamentary practice."[113] The centralizers' clauses were removed, and the bill passed.

Localism also dominated the turnpikes committee of 1840, which began its report by explaining that "no principle of common law is more clearly recognized than that which attaches to parishes the liability to repair all highways within their respective boundaries." The committee, which consisted of liberal MPs like the Duke of Richmond, Gascoigne Salisbury, Hatherton, Eliot, W. S. Lascelles, and Ayshford Sanford, voiced an intermediate course between previous attempts at rationalized centralization and the localist impulses now in the ascendant. The committee explained that this principle applied even to roads and streets that previous legislation had put under the care of centrally directed commissioners. It phrased this return to local practice in the language of public responsibility: "The obligation to maintain all public roads . . . is a public obligation, and in the nature of a public tax." Every parish, county, and turnpike trust had its "share of the public burden," "which is imposed for the general benefit of the community."[114]

The struggle between central and local powers came to a head again in 1856 when Lewis introduced a highway bill that proposed to "take away from all Parishes and Local Authorities the control over their own Taxation, and over the management of Highways, both of which they have always possessed." William Jolliffe struck him down by arguing that "they could not find a worse management for highways than a board of guardians."[115] The bills were opposed by William Hodgson Barrow of South

Nottinghamshire, Joseph Warner Henley, president of the Board of Trade, and Charles Newdigate Newdegate, the ultraconservative MP for Warwickshire, who were greeted as heroes by the localist press.[116] Editorials in the *Parliamentary Remembrancer*, Toulmin Smith's journal, argued that these MPs were defending the "Free Institutions which make the position of a true English Country Gentleman the proudest of the world." They aimed to defend those freedoms of all classes from elite privilege, the "class aggrandized at the expense of all the rest of the Community." In 1852, Barrow denounced Lewis's plans as "centralizing and bureaucratic, and as intended to take away from parishes the last remnant of local self-government." He accused the local boards of being a first step on the slippery slope of centralization: "How far were the Boards constituted under this Bill likely to appoint skilful officers? How long would it be possible for such Boards to go on without the control of some central authority? He believed not three years." He explained that "English roads" were the best in Europe, "the best in the world," and did not need further reforms for improvement. He thought that the present system of the "parish vestry" "accomplished all that could be wished for."[117]

Occasional proposals continued to suggest the centralization of the turnpikes, taking the same moderate tone of rationalized assimilation of parishes while preserving local authority.[118] These plans, whether nation based or county based, were consistently thrashed by localist opposition that viewed them as possible sources of interestedness and inefficiency. As different plans were mooted for the consolidation of turnpikes, localists consistently warned against any measure that would threaten the self-determination of the parish. Richard Milnes, the Chartist, warned his readers to guard against every opportunity for "taking the roads" from local into parliamentary hands. New measures threatened to render "this once happy land of freedom and liberty" a nation "covered with petty tyrants, which none could, or would control."[119] Toulmin Smith criticized countywide highway districts as a "means for fettering the action of those most immediately concerned in the condition of the Highways."[120] Others pointed to examples of failures in government arenas where turnpikes had been consolidated under central authorities.[121] Still others warned about "combination" in the consolidation of trusts, an opportunity for rank corruption unchecked by centralized propositions.[122]

The localist press equated government centralization with a conspiracy against the rights of freeborn Englishmen. The *Parliamentary Remem-*

brancer warned that "whatever Bill is introduced" in the future would "affect every Parish in England, and will need the most careful watching."[123] The paper labeled each bout of centralized legislation a "reckless contempt of all sound political, as well as economical principle," an "attack . . . on the institutions of England," the result of "irresponsible clique-government." It accused the centralizers of "the selfishness of the motives" and "ignorance of facts." The article jeered at the characterization of the bill as "reform," called it "the work of destroying Institutions," and accused it of taking power out of free parishes by a despotic seizure. Centralizing actions had to be guarded against by "private watchfulness," which alone could unveil "the true character of the attempt." Such language dogged the centralizing highways bill throughout its entire life. When it looked briefly like it might succeed, the *Remembrancer* announced that there had never been an event that could be "recorded with more shame, by whoever is capable of comprehending the spirit and the value of free institutions." Centralizers had the "sole object" of "bureaucratic ends," and each time these were defeated, the centralizers fastened them "on the end" again in new form. The paper rejoiced when the centralizers were "reduced to the humiliation of withdrawing the bill."[124] Still other publications took up the language of interest and conspiracy, demanding whether the common pedestrian needed any pavement besides "those green lanes and pleasant footpaths which have heretofore made the greatest charm of English country life."[125] They characterized any management measure as "trampling out the last spark of local self government, and the last vestige of English liberty" and "infusing poison into the lifeblood of the Constitution."[126]

Centralizers registered the shift in the winds as a definitive, political victory for localism. Alexander Mackinnon, MP for Lymington, another centralizer, conceded after their failure that the opposition derived from a bias against "the system of centralization" in general. He explained, "The country gentlemen of England felt averse to any plan of that kind, and they thought the power of superintendence and management should remain in their hands." Centralizing legislation would have breached the principle of local self-determination by uniting "wealthy trusts with poor ones," thus "saddling" the wealthy ones with "debts which others ought to pay."[127] Lewis was forced to admit "the difficulty of selecting a body to whom should be intrusted the management of highways."[128] Localist reporters triumphantly reported that the centralizers were "obliged to withdraw" the bill "as soon as its nature became known."[129] Well might Lord Brougham damn the localists as a bundle of "local interests and prejudices"; it was that movement that had won the day.[130] Indeed, localism's attitude of restraint on the part of central governance had set the tone for railroads, and it was localism that defined the map of the nation.

4

WAYFARING STRANGERS

Mobile Communities and the Death of Contact

> You know well how great is the difference between two companions lolling in a post-chaise, and two travellers plodding slowly along the road, side by side, each with his little knapsack of necessaries upon his shoulders. How much more of heart between the two latter.
>
> —WILLIAM WORDSWORTH

In 1791, two discharged soldiers were traveling from Portsmouth on their way home to Scotland. They planned to walk the entire route together, as soldiers often did, facing bullies and poverty as a team. When they crossed London Bridge, a stranger shouted at one of the pair, George Lowrie, who was still wearing his uniform. The stranger claimed that he too was a soldier and offered to join their company. A few nights later, when the soldiers were staying together at the Green Man at Poplar, their new friend robbed them of all their bundles, handkerchiefs, shirts, drawers, a psalter, and all. Watching from a nearby table when the bundles were snatched were two pensioners, one from the Thirty-ninth Regiment of Foot, another an officer. Together they hauled the thief before the Middlesex Jury and had him transported for seven years.[1] When one thinks about the vulnerabilities of strangers, this seems like a small punishment for the crime.

Despite the risks of cheats and frauds, most eighteenth-century travelers could afford to trust one another on the basis of small signs: the stranger's uniform signified shared time in the military, mutual acquaintances, and common experiences among a larger community of travelers. Soldiers remained vulnerable to individual cases of theft and malice, but a large community of travelers enforced the bonds of trust and punished deception.

Eighteenth-century highways were flooded with large numbers of travelers who moved in predictable patterns: soldiers rotating from point to point, journeymen tramping between stops on a circuit, or Methodists visiting coreligionists to study and preach. Members of these mobile communities expected to identify one another anywhere on the highway on the basis of a few cues. During the eighteenth century, those who identified as travelers carved sanctuaries of solidarity out of highways where strangers met one another only by chance.

The new hordes of travelers were often, like the roads themselves, a consequence of the state's expanding shadow. From the 1690s onward, Britain's military-fiscal state was already sending travelers out from the capital onto the circuits and rotations of officials. Excisemen, postboys, surveyors, and soldiers followed the roads and highways of Britain, collecting funds to build the navy, carrying official instructions, tracing the shape of future highways, and policing the rebels of the Scottish north. In sheer numbers, the greatest presence was the military. By the 1730s and 1740s, large numbers of soldiers were circulating in packs over the British countryside: regiments billeted in market towns, recruiting parties riding into town, and discharged soldiers tramping in search of work. In an age when few individuals traveled beyond the realm of their village, soldiers learned to identify and depend on one another in times of need. Individuals who identified with this community asked other such travelers for directions, material support, friendship, and defense. Throughout the eighteenth century, groups like Methodists and artisans adapted the military system of highway rotation to their own purposes: sharing enthusiasm deep across the country or securing the needs of unemployed journeymen.

Individuals who belonged to these communities could confidently travel hundreds of miles in search of new places to preach or work, knowing that wherever they went, other travelers would identify them and welcome them home. Bonds of trust were enforced by the vulnerability of travelers to frauds and the resistance such travelers faced in towns. Travelers turned to one another as sources of identity, confidence, and community: Methodists greeted one another as "fellow pilgrims" and artisans as "fellow travelers." They reinforced the ties of identity with shared songs, signs, and secret handshakes. Unified by their common vulnerability, these communities gained confidence in their identity as a people united by road travel. For those who shared an identity with other travelers, the highway of strangers was dotted with communities of friends.

In this chapter, I turn from the management of roads to their use in everyday life. We scrutinize travelers' personal narratives to determine the texture of trust on the road. This texture changed with the introduction of the post coach in 1784, when middling travelers—these town shopkeepers and business travelers did not yet share a national identity that would bind them together into a middle "class"—began to travel the highways in larger numbers. The middling travelers negotiated a new world of travel, characterized by consumption: using inns, coaches, and guidebooks, they found their way from place to place, cushioned within a new set of tools for navigation. This world of consumption, I argue, isolated middle-class travelers from the dependence on strangers that had characterized earlier forms of travel. Freed from the logic of need, middle-class travelers developed new strategies like "small talk" to filter their contact with the other travelers in the same stagecoach. Indeed, even as the middle class imagined a world of respectable travel, parliamentary highway builders were arguing that roads were a major force of civilization, an agent for shaping model citizens. Their arguments and the fears of middle-class travelers played out after 1822 in an onslaught of legislation designed to police the mobile communities whose structures had once defined sociable meetings between strangers. Criminalized as vagrants, the mobile communities were dispersed, and nineteenth-century travel was reshaped as a zone of middle-class respectability that replaced the former pattern of open exchanges with an attenuated system in which middle-class travelers could avoid strangers altogether.

Soldiers: Traveling for the Military-Fiscal State

Waves of recruitment at the commencement of each stage of hostilities produced a general upward trend in army size during the eighteenth century.[2] The army's average annual enrollment numbered 76,404 at the Nine Years' War, 108,404 at the American War of Independence, and 300,000 in 1809. Having increased over the century, military recruitment reached a peak with the Napoleonic Wars, with over 75,000 recruits deployed between 1803 and 1815.[3] Each phase represented movement of massive numbers of humans across the nation. Individuals recruited or impressed from British villages, towns, and cities circulated through the roads to the ports, leaving for the Continent and points beyond, at the beginning of each war. At the close of each campaign, soldiers returned to Britain in droves, traveling back from the ports to the towns and hoping to find the makings of a living along the way.

The system of rotations and deployments plucked individual soldiers from the confined horizons of village life and thrust them into a world of frequent travel. Samuel Hutton fled to the army in the 1740s when he broke his apprentice contract. He immediately entered a regime of intense travel. His first march, between Nottingham and Derby, took him one hundred miles by foot. He had been a skinny and sickly child, but the march and rations together changed his physiognomy and health, and he grew four inches in the next six months. In the next year, he marched from Derby to Scotland, boarding in small towns all the way.[4] Benjamin Harris, a Dorset rifleman, had never left the fields bordering Stalybridge where his family had been shepherds till 1803, when he was drafted. He spent three months training in Winchester, journeyed with his regiment to Portsmouth to witness the execution of a deserter (where he met soldiers from the Isle of Wight, Chichester, Gosport, and elsewhere), and joined a recruiting mission in Cork, Dublin, Cashel, and Clonmel before returning to England. The regiment landed at Pill and marched to Bristol, Bath, Andover, and Salisbury Plain (where the Irish caused a riot) and thence to Ashford, in Kent, where Harris joined the riflemen. All these travels took three years. From 1806 to 1810, Harris rotated through the Atlantic circuit: Denmark, Obidos and Rolica, Vimeiro, Lisbon, Salamanca, Sahagun, Corunna, Vigo, and Walcheren. His companions during these rotations were men from Leicestershire, Nottingham, Yorkshire, the Highlands, and Dublin.[5]

Military rotation required frequent movement and extensive travel through both Britain and the larger empire. About half the army at any given time was stationed in Britain itself. Which soldiers remained in Britain depended on what position in the army they had taken. During peacetime, one-quarter of the foot soldiers remained on foreign garrison duty, half in Ireland, and the rest in Britain.[6] The Horse and Guards were stationed permanently in the British Isles, deployed abroad only as required for action; during peacetime, they remained at garrisons scattered throughout England and Scotland. Others rotated between Britain and the empire, as they did among England, Scotland, and Ireland, through cycles of both peace and war.

As they embarked and returned, these troops passed through a constellation of common towns, where they might stay and mingle. Those transported by sea entered and left at Plymouth, Portsmouth, and Bristol, or occasionally Dee, Liverpool, Newcastle, Chatham, Rochester, and the Clyde ports, or the Thames ports of Greenwich, Woolwich, Dartford, or

Gravesend. Those going to Scotland passed through Carlisle, Berwick, and Coldstream.[7] Military rotation placed a significant proportion of occupying troops inside Scotland and Ireland as well.[8] Such depots were physically organized to accommodate large numbers of comers and goers. Portsmouth had four good barracks: two for the invalid regiment in garrison; a third for the companies of artillery, who do duty here; and the fourth for the marines of this division, "that they may be in readiness to embark on board the ships as they are wanted."[9] Such barracks and billeting towns throughout the country were positioned so as to make mobility instantaneous when required. Quarters at Eling and Newport near Southampton stood "ready, on any emergency" to deploy any one of the "various regiments [that had] successively occupied its barracks . . . to march round to Southampton and embark for foreign service."[10] Military labor and the military on rotation appeared, in each of these centers, as a honeycomb of separate workshops manned by constantly circulating human hordes. Portsmouth, which had been a military arsenal since the 1720s, had at any time "three regiments of infantry, a division of royal marines and marine artillery, detachments of royal artillery, and engineers for repairing the work."[11] Even Yarmouth, a minor port, could amass a daily parade of "sailors walking about, and the carts jingling up and down over the stony lanes . . . past gasworks, rope-walks, boatbuilders' yards, shipwrights' yards, ship-breakers' yards, caulkers' yards, riggers' lofts, smiths' forges, and a great litter of such places."[12] The movements of soldiers formed a major stream of traffic in port towns across the nation.

The volatile pattern of army expansion gave large numbers of Britons some experience of military life.[13] The army recruited gentleman officers, professional artisans, harvest laborers, and the very poor, incorporating all classes of the British into the geographic pattern of military life.[14] As a result of these tours, rapid and widespread, recruits were drafted from every quarter of the four kingdoms, with an increasing burden on Ireland and Scotland as the period progressed. Irish and Scots thus joined Englishmen rich and poor on their rotations through the ports, towns, and villages, creating a new class of outsiders circulating through the English streets.[15]

Methodists: Emulating the Military-Fiscal State

Because of their alienation from local communities, their frequent travel, and the hardship of their life, career soldiers became obvious targets of Methodist conversion. Methodist preachers, quick to notice the isolation of soldiers, learned to capitalize on the ready-made community of potential converts to extend the geographic reach of their evangelism. Ministry to the army doubled for Methodist preachers as a source of converts and a vehicle for further evangelism.[16]

Children of soldiers who were frequently on rotation converted to Methodism in high numbers. Samuel Bradburn, afterward a famous preacher in Manchester, was a soldier's son who had met the Methodists who fought in the battle of Dettingen.[17] Caroline Hopwood, a Methodist convert and later the Quaker leader in the antislavery petitioning movement of the 1770s, was the youngest child of an impoverished Scottish army lieutenant; both her father and her brother died in the army while abroad before she was ten.[18] Such individuals had been raised among the constant circulation of army families.

The early chronicles of Methodism lay bare its close relationship to the military. First, soldiers were the other major group that was already as widely dispersed as the itineraries that ambitious missionaries had given themselves: George Whitefield preached to the soldiers who were on his ship in 1738 as he sailed to Georgia.[19] He happened to find colonial troops in Boston when he was a missionary there in 1745, and he preached to them too.[20] Eighteenth-century outposts of empire like Halifax, Antigua, Philadelphia, and Gibraltar became the territories targeted by evangelists. There early missionaries like William Black and Freeborn Garrettson reached settlements where the unchurched or badly churched leaped to embrace the warmer passions expressed by preaching Methodists.[21] Desperate for sympathy as well as redemption, the military converts met "for prayer and conversation" wherever they could, hiding from the regiment "in dens and caves of the rock."[22]

Second, Methodists became the particular targets of press-gangs wherever preaching raised local outrage, so the ranks of the armed forces were dotted with Methodists who had been taken prisoner. Congregationalist preachers like Matthias Maurice and Abraham Gill, missionaries to South Wales and Lincolnshire, were victims of naval impressment in the first two decades of the eighteenth century.[23] James Ingram, who belonged to preacher Howell Harris's Welsh militia, was seized by a press-gang in 1744 and preached three times a day to the others who had been seized by the press-gang.[24] Dissenters in Suffolk in the 1810s had their meeting house smeared with human feces and were threatened with impressment if they stayed.[25] Such terror tactics had the unlooked-for consequence of cramming the ranks of the military with evangelists.[26] They also succeeded in disseminating Methodism to the far corners of nation and empire.

Third, Methodist leaders had begun to recognize the geographically dispersed, disenfranchised body of soldiers as an ideally fertile ground for their conversions. John Wesley's experiences among soldiers made an early mark on his sympathetic conscience, and he took particular pains to preach to them throughout his travels.[27] Other preachers dedicated the whole of their career to evangelizing the military. George Whitefield was a soldier at the time of his conversion and did his first preaching during his military rotation in Gloucester, Stonehouse, Bath, Bristol, and Gibraltar.[28] Howell Harris, the "soldier preacher," emerged as an independent field preacher in Wales in 1735, about the same time as Whitefield.

Harris pioneered the "highway pattern" of proclaiming the gospel as an itinerant and was followed by Daniel Rowlands, Whitefield, and Wesley.[29] As he became more cognizant of competition among the famous preachers for fame and geographic spread, Harris traveled between Wales and London with great frequency. In prayer and in his personal journals, he struggled against jealousy at the greater fame of Whitefield and Wesley. His solution was to leave the overpreached fields around London in order to concentrate on ministering to the military.

In 1759, Harris joined the Brecknockshire Militia. Traveling to Yarmouth, Brecon, Abergavenny, Torrington, Bideford, Barnstaple, Bridgewater, and Plymouth, he took the opportunity of his battalion's movements to preach in towns and villages throughout the southwest, simultaneously charged as Methodist missionary and captain of a regiment of Welsh Fencibles.[30] Harris, preaching "in his regimentals," succeeded "without much opposition" where earlier civilian circuit riders like Thomas Olivers had been chased off by the mob.[31] Wesley encountered Harris's work when he visited Yarmouth in 1761 and was astounded to find the port town morally transfigured from a zone of "wickedness and ignorance" as bad as "any seaport in England" to an exemplary mission populated by evangelical soldiers. "Some of them now earnestly invited *me* to come over," Wesley remembered.[32] Harris records converting many of the men in his militia, lecturing them virtually every night on the gospel and the soul.

Much of the conversion activity took place during the European wars of the 1740s and 1750s, where soldiers encountered one another far from home in harrowing circumstances. Methodist soldiers' autobiographies rely for their melodrama on narrow escape from battles where officers' heads were blown off by cannonballs and dying friends weeping in each other's arms commended each other to the Lord. Methodist conversion emphasized this personal narrative of drama, choice, revelation, and salvation; the heavy campaigns of Dettingen and Fontenoy provided an ideal backdrop. Later preachers enjoyed the same pattern of conversions whenever they reached out to the army. Administering a new circuit around Glasgow in the 1780s, John Pawson corresponded with other itinerants about the remarkable enthusiasm of English soldiers stationed there, his best hope in countering the "Calvanistical devil" and irreligion otherwise prevalent in the town.[33] Gideon Ouseley, the rural revivalist of Ireland in the 1820s, was converted by a Methodist quartermaster in the Royal Irish Dragoons whom he met at the inn near the barracks in Dunmore.[34]

Circulating soldiers required cultural and religious forms suited to far-traveling, easily dispersed bands of brothers. Military travel was enough to put many soldiers in the way of Methodist teachings; a soldier of religious inclination like John Haime would have trouble developing prayerful relationships with local churches when he was moved around so frequently, but he could develop stronger relationships in the company of other soldiers. Haime, a gardener and button maker of Shaftesbury, was driven by severe fits of melancholy to abandon his first wife and enter the dragoons. He went to church in every town where the regiment stopped, but he swung rapidly between ecstasy and dejection, isolated from anyone else who understood the severity of his experience. His conversion was assured only when he met one of Whitefield's preachers in London, where he had gone with a camp equipage. He encountered Charles Wesley in Colchester and spoke to him at length after the service. When the regiment left for Flanders, he corresponded with Wesley at great length and began preaching. Amid the dead bodies floating on the Maine, he and other soldiers formed a prayer group. "Hereby new love and zeal were kindled in us all," he recalled. "Was there ever so great a work before, in so abandoned an army?"[35]

The horror of military life drove soldiers to religion. Haime's depressive rages against the devil took aggressive forms. When he came across a soldier who was expressing relief at finding a lost coin, for instance, Haime took the opportunity to bicker with the man's claims to being lucky. The lost coin came as much from the devil as from God, reasoned Haime, who saw temptation everywhere. The exasperated man protested, the two got into a fight, and Haime tried to have his would-be convert court-martialed. Such shows of passion as these were not uncommon. In the camps, nervous soldiers huddled far from home in the midst of bloody battles where Haime's friends and officers perished in large numbers. In an atmosphere of dread, demonstrations of passion could woo the hearts of lost soldiers. In his time on the continent, Haime converted six preachers and drew a society of three hundred. By the time of Haime's first camp meeting, word had spread. A thousand soldiers came, both officers and regulars. He had General Ponsonby's assistance in using the English church in Bruges for a congregation before the battle of Fontenoy. After returning from the war, he was discharged, was taken by John Wesley as a traveling preacher, and retired to St. Ives in 1766.[36] In terrifying circumstances, where soldiers were disconnected from traditional communities of support, Methodist fellowship offered a refuge of community, hope, and comfort.

Methodist links were strongest in maritime and military enclaves. Captain Webb, a retired army officer, started preaching in uniform at Winchester in 1783; soldiers flocked to hear him, and he eventually converted his commanding officer, Corporal Thomas Miller, who was later active in persuading Wesley to send bilingual preachers like Robert Carr Brackenbury to the Channel Islands.[37] As a result, the earliest Wesleyan itinerants in the Channel Islands were military or military affiliates in high numbers. This is hardly surprising, given that the army barracks there provided the most important link between the islands and the mainland. Captain John Brown of Poole was responsible for transporting cattle for the victualing of the garrison; he began preaching there in 1779.[38] The Methodists directly capitalized on military patterns of social geography.

The Methodist affair with the army lasted until the Napoleonic Wars, when Wesleyanism's audience began to shift to the settled, town-dwelling middle class. For most of the eighteenth century, however, Methodism followed the shape of circulation set in gear by the military-fiscal state. It established mobility not only as a professional identity, but also as a spiritual and cultural one.

Methodism adapted the military system of rotation to its own purpose and so launched a wave of traveling preachers whose identity was shaped by the experience of a community most visible on the nation's highways. Methodist polity espoused a mobile and open community over any local one.

Early Methodists rejected built meetinghouses, preferring public sites for their ministry, where they could attract any who would come. George Whitefield and John Wesley pioneered "field preaching" in the 1730s, but "field preaching" was technically the American term; in Britain, "street preaching" was preferred and was more descriptive of the particular attitude toward public space it entailed. "Street preaching," by stopping up the ways with crowds, formed an offense under the Riot Act. Numerous preachers were arrested for street preaching, which gave them the attention of diverse passersby but put them, in the eyes of the law, on the level of common hustlers.[39] Town resistance to Methodism made street preaching dangerous. When Howell Harris stayed in Machynlleth, the mob rioted, throwing stones and firing pistols until he stopped. He avoided harm by preaching to the street from windows in the inn or house where he stayed.[40] Determined preachers would do whatever was necessary to make their voices heard in the street.

Methodists early called themselves a "connexion" rather than a community.[41] Richard Cawley, a correspondent of Wesley's in the 1740s, stated, among the early rules of his Cheshire meeting, "We have nothing to the News of the Town, and of the Business of others: But we desire to hear of Things pertaining to the Kingdom of GOD."[42] Biblical language offered the shape of the kingdom of pilgrim preachers as an alternative to the political boundaries of the town. Converting the language of the Psalms to contemporary experience, Methodism offered a communal identity unified by mobility.

Methodist leaders exploited the flexibility of their emerging organization to transform preaching into a tool for reaching the other mobile people of Britain. Evangelizing travelers served a double aim: they gave Methodists access to displaced persons like soldiers who desperately needed a community, and they put Methodists in the way of the diverse cross section of the road's users. Preachers in such circumstances persisted in street preaching because it offered an interface with the public unavailable to those locked in antique buildings. In 1814, Mr. Chater, a Baptist preacher, gazed from his window at "the numbers passing and repassing" that had never heard "the word of life" and longed to become a street preacher.[43] Preachers considered the roads to be avenues crammed with potential converts waiting for their witness. Devoted itinerant preachers took every one of their journeys as an opportunity to evangelize. Adam Clarke, on his way to Kingswood for the first time, took the opportunity of the stagecoach ride to proselytize, and "talked divinity and quoted Horace" the length of the journey, to the other passengers' "no small surprise and amusement."[44] Wesley too boasted of chance converts recruited from conversation in post chaises.[45] William Bramwell, an energetic advocate of Methodism in the

Fylde during the 1780s, converted Ann Cutler, the poor hand-loom weaver and future prophet "Praying Nanny," when they began talking as they walked between Longridge and Ribchester.[46] William Carvasso of Mousehole, who evangelized throughout Cornwall in the 1780s, buttonholed everyone he met on the road, from a tollhouse keeper's daughter to a dandy riding above a coach.[47] Street evangelism aimed to make the preacher unavoidable, and as a result, the Methodist presence was a constant feature of highway travel. The preacher Cornelius Cayley, riding on circuit between meetings that he would lead, passed another preacher, John Cennick, "preaching in the high road."[48] Methodists exploited the road network and used their everyday journeys to capture attention.

Mobility informed the political structure of Methodist circuits. The annual Methodist Conference of preachers was established in 1744, with its first meeting in London. The plan held that the conference would circulate every year among Newcastle, Bristol, and London, thus making a geographic circuit of England in much the same way as the Quaker circuits had for years. Circuits were appointed from 1746 onward, following from initial earlier traveling plans of Methodist pioneers William Grimshaw and John Bennet, early converts who in the 1740s preached in circuits around Manchester. A geographic concentration around the south followed from the Wesleys' initial efforts around Bristol, Kingswood, and London. The first seven circuits of regional preaching rounds were established in 1746, namely, London, Bristol, Cornwall, Evesham, Yorkshire, Newcastle, and Wales. From 1738 onward, pioneers like Benjamin Ingham (a colleague of Wesley's from Oxford) and David Taylor (a protégé of Lady Huntington) pushed Methodism north to Leicester, Yorkshire, Lancashire, Cheshire, and Derbyshire. Itinerants went to Ireland in 1747 and Scotland in 1751.[49] By the 1750s, there were thirty-six preachers in England, Wales, and Ireland, preaching on nine circuits and serving almost a hundred societies, most concentrated in the three strong southern circuits west of the Thames along the Bristol Channel.[50] These preachers were further formalized into the first formal circuit plan in the 1754 conference, which outlined daily routines of preaching for each preacher. By 1765, there were thirty circuits in Britain, with a total of twenty thousand attendees.[51] By 1791, when John Wesley died, there were seventy-two circuits in England, twenty-eight in Ireland, seven in Scotland, and three in Wales, each attended by several lay preachers who moved regularly from place to place.[52]

The Methodist connection was therefore an organization composed of, and run by, itinerants. Wesley's own travels were prodigious. In 1787, when he was eighty-four, a typical week records preaching at the annual conference; traveling with fourteen preachers by coach from Manchester to Bristol (with two breakdowns, the journey lasted nineteen hours) and preaching immediately on arrival there; then going by coach to Gloucester (eleven hours) to deliver an evening sermon; then to Salisbury (fourteen hours, leaving at 2 A.M.); then to Southampton (4 A.M. departure), preaching twice; and then by sloop to Guernsey.[53] William Thompson, the first president of the conference after Wesley's death, was an itinerant for forty years.[54] The church was divided into smaller conferences and circuits, each of which was overseen by an itinerant preacher who spent the majority of his week traveling during the day and preaching in the evening.

Traveling preachers were usually stationed at a particular circuit for a short period and then were rotated around the country. Even local preachers were expected to keep on the road. Alexander Menhinick of Sladesbridge, a shopkeeper and local preacher in Cornwall, traveled twenty miles by turnpike every Sunday on the Camelford Circuit.[55] These distances were extremely arduous. The Cheshire Circuit given to John Bennet included Lancashire, Nottinghamshire, Derbyshire, and parts of Yorkshire as far as Sheffield.[56] In the 1750s, the Haworth Circuit, supervised by two traveling preachers, stretched 120 miles long and 60 wide; even so, they managed to hit Preston every six weeks.[57] Adam Clarke's Norwich Circuit included Norwich, Thurne, Yarmouth, Lowestoft, Cove, Beccles, Wheatacre, Haddiscoe, Thurlton, Heckinham, Hempnall, Loddon, Barford, Hardwick, Stratton, Tasburgh, Dickleburgh, Winfarthing, North and South Lopham, and Diss. Clarke remembered, "It cost us about two hundred and fifty miles a month; and I have walked this with my saddle-bags upon my back."[58] The ideology of travel as evangelism made itinerants all the prouder to ride when they faced hard weather. The biographer of one such preacher explained, "We therefore find him braving the storms and tempests, from one place to another, traveling on foot through snow and mud, where the roads were too bad to admit his traveling on horseback, that he might, as widely as possible, extend the empire of his divine Lord and Master."[59] Prodigious travel characterized the whole of the preacher's life. It was not uncommon for preachers to boast forty years of itinerancy, stopping only when physical breakdown made the process impossible. John Murlin, an itinerant from 1754 to 1787, stopped circuit riding only "when, through heavy affliction, he was obliged to become local."[60]

Although Methodists remained only about 1 percent of the population for most of the eighteenth century, by the 1780s their growth became exponential. Between 1741 and 1765, two hundred Methodist itinerant preachers were accepted. The majority returned to serve as local preachers or left for the established or dissenting churches with their static parishes; only eighty-one died in the work.[61] Between 1770 and 1780, British membership doubled from two thousand to four thousand. By 1810 it had exploded to fourteen thousand and by 1840 to twenty-six thousand. This central core of formal members was surrounded by a wider circle of service attendees and occasional visitors.[62] The geographic reach of these conversions paved the ground for other, later developments that spread through the coherent, centralized structure of the Methodist Church, for instance, the Sunday schools movement, so successful that by 1851, two million children in England, or 75 percent, had some experience of Methodist Sunday schools.[63]

Tradesmen traveling in search of work encountered itinerant preachers when they met one another on the highway. John Cennick, one of Wesley's first teachers at Kingswood, had run away from apprenticeship no less than eight times as a carpenter and for other trades in the 1730s before he reached a state of despair. Apprenticeship took him from Reading to London and initiated him into the life of a traveler. When he converted, he was working as a surveyor, traveling between the estates of gentlemen, and in this capacity he made his way to Oxford with the intention of looking up Charles Kinchin, one of the preachers of whom he had heard so much.[64] Alexander Mather, the famous preacher, met Wesley in London, where he had journeyed from Brechin in Scotland as a journeyman baker in search of work.[65] Around 1750, Thomas Olivers was converted from a life of traveling crime and poverty by a Methodist journeyman apothecary who found him in Wrexham.[66] John Bennet, one of Wesley's early converts and among the first evangelists to Cheshire, was invited to Chester by a tailor, George Shaw, a resident of Boughton, who had heard preachers in Bolton and elsewhere.[67] John Nelson of Bristall, son of a mason and a traveling stonemason himself in the 1740s (the masons were among the most mobile of the trades), was taken by business to London, where he encountered the Wesleys in 1739; he encountered John Wesley again in 1742 when they were both in Newcastle.[68] Nelson's influence would allow

Wesley's Oxford-trained friend Benjamin Ingham to found the forty Methodist societies that sprang up after 1742 in Yorkshire.[69] A Methodist sea captain named Joseph Turner of Bristol brought Methodism to Cornwall when he founded a religious society at St. Ives in 1743.[70] William Bramwell, a native of Elswick, traveled as an apprentice currier and fellmonger to Preston, where he encountered Methodists; in the 1780s he would become one of the most famous preachers of the Fylde, traveling by horse on a journeyman's wages.[71] Samuel Drew, the "metaphysical shoemaker" of Cornwall in the first decades of the nineteenth century, trained as a cobbler and a printer in London and Liverpool before he returned to Cornwall as a local preacher.[72] Mr. Tripp, one of the local lay preachers around Yarmouth in the 1780s and a correspondent of Wesley's, was a cooper and a fishmonger in one of the military towns where Methodist converts were coming and going.[73] Methodism ministered to the isolated; early Methodist preachers traveled long distances to remote communities and encountered similar traveling populations, separated from their roots and from other religious communities. The changing demographics of Methodist converts reflect the changing population on British roads at any given time.

Domestic evangelism on the highway was supplemented by the international networks to which soldiers belonged. The Methodist ideology of publicity, linked to the travels of soldiers, resulted in an even wider range of influence. In Manchester in 1754, Wesley encountered seventeen dragoons who had been in Haime's regiment in Flanders. All had rejected Haime at the time, but they found themselves strangers back in Manchester. Then "one and another dropt in, he scarce knew why, to hear the Preaching. And they now are a Pattern of all Seriousness, Zeal, and all holy Conversation."[74] Some missionaries went even farther. John Baxter, a shipwright at Chatham Dock, was one of Wesley's early preachers and became a missionary to Antigua when he was sent there by the navy.[75] Captain Thomas Webb was sent to the New World with General Wolfe at the conquest of Quebec in 1759. After losing an eye and returning to England, he heard Wesley preach in Bristol and converted. He preached in Bath and later in Albany, New York, where he was barrack master after 1765. From there he evangelized, "all life and fire," in Philadelphia, Long Island, Dublin, Winchester, and London.[76] Methodism's vast international spread reflected a community of converts who were already travelers by the nature of their profession.

By the end of the century, Methodist itinerancy was taking on a new form. Older itinerants like John Pawson complained about the lax habits into which new preachers fell. They looked to the longer circuits of the 1740s and 1750s as a model for active ministry that had been discarded in an age of better-funded churches.[77] Pawson complained bitterly about well-funded Methodist parsons who took their wives with them in post carriages, when he himself had walked the length of Britain for many years.[78] Itinerancy was carried forward by split sects like the Primitives and evangelicals. William Garner of Hull told the Primitives of Hull, when he retired, that he had traveled continuously for twenty-one years between 1821 and 1842, reckoning that he had gone "on foot, with comparatively trifling exceptions, 44,936 miles," or about "twenty and (occasionally) thirty miles a day," not including "daily perambulations in the cities, towns, villages &c. where my lot has been cast," in order to preach "6,278 sermons," not including "exhortations, addresses, missionary speeches &c." He reckoned his prodigious feats relatively average for the community.[79]

Methodism was forged on the mobility afforded by roads and in the mobile communities it encountered there. Goaded by evangelism and the precedent of military travelers, Methodist preachers pioneered the shape of a larger community united by travel. Such a mobile organization allowed the connection to flourish and established an important precedent for patterns of community and trust among the mobile poor.

Artisans

Journeymen and tradesmen floated in and out of the army, traveling with it across the nation and the empire. Therefore, they also became targets of Methodist conversion and developed opportunities to learn the tactics of travel from both camps. Military experience inspired artisans with the confidence to travel to unknown territories in search of work.

During the eighteenth century, apprenticeship, traditionally one of the most secure means of finding a living, had become among the least secure.[80] Army recruits were frequently gathered from the ranks of apprentices and journeymen who were most vulnerable to patterns of economic flux.[81] Samuel Hutton was one such recruit who found the army and its geographic mobility a positive substitute for another, less stable form of mobility. The son of a full-time drunk who whipped him regularly, Hutton worked in the silk mills of Derby from age seven. In 1743, at age ten, he was sent to Nottingham as an apprentice to his stocking-maker uncle. Trapped in an outdated system, Hutton, like many others, was a victim of history. He tried several times to escape (a common tale among apprentices of the era, who could legally be captured and returned to their masters under contract) and became a mobile recluse. Beggars shared their food with him, and he became the object of charity from highwaymen. In his autobiography, Hutton summed up his experience: "There was but one asylum before me—the army."[82] In contrast, professional artisans who had already made their place in a guild might join the army only if they encountered hard circumstances. John Brown, a trained shoemaker, was recruited to the army only after tramping for months in an unsuccessful search for work.[83] Joining the army gave artisans, as individuals, a source of economic security and exposed them to the structures of modern travel.

Former soldiers, accustomed to travel, became artisans after training in the army. Benjamin Harris, the Dorset rifleman, became his regiment's shoemaker before he returned wounded and set up a cobbler's shop in Richmond Street, Soho, London, and he mentioned other soldiers who became bandmasters or cooks while abroad.[84] Learning a trade could help the transition back to civilian life, always difficult for worn-out, infirm, or wounded soldiers with their minimal allowances. Such army-trained artisans brought confidence in the solidarities of mobile peoples to the trades and offered their brother tradesmen a particularly deep knowledge of dealing with far-away communities, which benefited from and enhanced the loyalties of craft. Tramping stonemasons in the 1840s, looking for work and a place to stay, met fellow artisans who had joined the militia, who smuggled them into the barracks for the night.

Such opportunities to learn the skills of traveling were redoubled by artisans' experiences with Methodism. Circulating through the army and the roads, artisans necessarily came across Methodists. Tradesmen were particularly likely to encounter Methodist preachers on their mutual journeys, and journeymen often established a familiarity with Methodists through frequent encounters on the road. While on the tramp in Gloucester in the 1840s, boilermaker Alfred Ireson met other Wesleyans who lodged him for his stay.[85] Affiliation with Methodism gave artisans an identity inside a mobile community that could cushion their further travels. By the 1770s, journeymen experienced in military rotation and Methodist circuits applied the structure of such mobile communities to their trades. Combinations of well-organized journeymen, capable of setting prices and standards against their masters, began to appear by the 1750s. These "labour exchanges" appeared in the inns and public houses in which employers typically went looking for their short-term labor needs.[86]

The houses of call were also typically the first place a journeyman on the road would look for employment when he arrived in a new town, and geographically dispersed journeymen began to share personal maps of towns, public houses, and possible employers. Journeymen connected by these centers gained the confidence to travel long distances with the expectation of support from a larger community.

The ritual of "tramping" began to be organized, trade by trade, from the 1740s onward.[87] Faced with economic pressure, new trades adopted the system of rotation to their own ends. Weavers and shoemakers, whose trades were dependent on highly variable employment conditions, were among the first trades to mobilize and organize informal associations on the road. Shoemakers in particular were associated with disconnection from their communities and with radical organizing.[88] These trade-by-trade associations were gradually replaced by societies that catered to the mobile trades as a whole. The Oddfellows, first appearing in 1748 in Southwark and Smithfield, supplied "Grand Circuit Quarterage," which provided the cost of bed and board per diem to members who needed employment.[89] John Brown described how the system worked in the 1790s. While costs drove bread "up to famine price," journeymen were laid off in large numbers. Fortunately, the trades had saved tramping money to send them off for work: "Nearly all the single men packed up their kits and went on tramp, as they were not allowed any relief; but I remained in town, and determined to pledge my clothe, books, and such valuables as I possessed."[90] The average distance of a single tramp to find work could mount to 334 miles in certain trades, and single individuals walked even further. In a single year, boilermaker Thomas Watson walked 1,332 miles, rotating among Hull, London, Manchester, Newcastle, Southampton, and Bristol.[91] Formalized associations gave artisans the same security in travel enjoyed by soldiers and Methodists.

These associations flourished in the hardship of the Napoleonic Wars, an era when out-of-work tradesmen like John Brown could avoid starvation by relying on tramping routes. By 1808, the carpenters of Birmingham had built a house for "the reception of working men traveling for the purpose of getting employment, and who are commonly called tramps."[92] Oddfellows' lodges had spread to Manchester, Salford, and London by 1810 and systematically offered their out-of-work members the cost of a day's bed and board when they arrived, and the next day's bed and board if work could not be found.[93] Deliberate attempts at organizing a national organization spread quickly. By the 1820s, links existed within at least twenty-eight trades. Humphrey Southall suggests that the movement reached its apex in the 1830s and 1840s, when mechanics could offer one another a clubhouse in every town.[94] The size of the national organizations was prodigious. By 1815, Lancashire had 147,029 members of friendly societies. In 1821, London had 119,498 out of a population of 1,144,531. High numbers prevailed among carpenters, shoemakers, blacksmiths, and tailors, all known as tramping trades.[95]

The affiliated friendly societies based the shape of their organizations on the rotating circuits of Methodists and soldiers. The Oddfellows can again serve as a meaningful example. The first meetings—the Loyal Abercrombie, the British Volunteer, and the Victory Lodge—referenced the recent wars in which many tradesmen had seen action. By 1826, the Oddfellows had a Grand Annual Moveable Committee that rotated, like the Methodist Conference, through the major points of the country. Its first meeting was held at the Prince's Tavern in Manchester.[96] Its form of organization ensured similarity of approach throughout the nation, and the order began its most rapid phase of expansion after 1832. In 1833, the Nelson Lodge, Kendall, built the first "permanent hall" in the order (previous meetings had been held at inns and theaters). In 1837, the first Scottish lodge emerged in Aberdeen, founded by wool combers who had recently migrated from Yorkshire.

Artisan organizations supplemented the community Methodism offered to mobile individuals. The identities they both offered were not exclusive, however. In the 1810s, Robert Pilkington of Bury, J. Grundy of Manchester, and Samuel Bamford served as both union organizers and Methodist preachers or schoolmasters. The pattern of overlap and interdependency continued over the decades that followed.[97] Military rotation, Methodism, and artisan organizing formed overlapping systems of solidarity among strangers that gave individuals the confidence to travel great distances, sure of sources of support. Mobility was a fact of life for many individuals in eighteenth-century Britain, but its nature was multiple and variable, open to a wide range of formal and informal associations in which individuals could participate during the course of their lives.

Other Migrant Laborers

Unlike soldiers, Methodists, or artisans, the unskilled poor remained outside most of the eighteenth-century rush of organization. The Salopian Amicable Society appeared in Manchester in 1785 for natives of Shropshire who had migrated the short distance in search of work, and Scots societies for migrants appeared in London and Norwich in the 1770s.[98] However, dues of friendly societies remained above the short-term prospects of most of the mobile poor.[99] General trends of migration remained constant from the beginning of relatively urban times till the nineteenth century: historians of migration have proved that throughout all modern periods, the great majority of individuals moved only within adjoining parishes during their lifetimes.[100] Most of these migrants traveled for seasonal work in the building trades or followed the harvest.[101]

On the footpaths were porters, laborers, and tradesmen, identifiable by their bundles, which reminded James O'Flanagan, an Irish law student, of "an ant-hill in full work, each having a separate store, hurrying to that, laden with the produce of his industry, or leaving it to dispose of the result of his labour."[102] Their purpose could be further distinguished by what they carried. "Men and women with fish-baskets on their heads" were moving toward Billingsgate; artisans on tramp carried their few possessions in "a kind of wallet on their backs."[103] This "unbroken concourse of people trudging out with various supplies" stayed clear of the carriageway, heading by foot toward markets while carriages rolled swiftly toward their official destinations.[104] Travelers of all classes banded together for safety and warned one another of the dangers of certain towns.

By 1800, these informal networks constituted a majority of those on the road, defined by journalist Henry Mayhew in the 1850s as the "wandering tribes" that characterized British life on the highways and streets of the metropolis. The first groups that supplemented the traveling ranks of soldiers were the empire's peripheral and deracinated poor. Like the soldiers, they too were driven by the activities of the military-fiscal state. Scots poor displaced by famine tramped southward after the 1750s. The Irish poor followed, likewise driven by wartime famine, after 1790.[105] In the peace after 1815, economic stagnation in Ireland brought floods of migrant workers to England; English parishes, no longer willing to support the burden in any form, refused to relieve them, and a new tide of circulating poor was thus created. By the 1840s, 105,916 persons, or 6.3 percent of the population of Lancashire, had been born in Ireland; one-third of the population of Liverpool and nearly all the dockworkers of Liverpool were Irish. Of these, large numbers traveled to the south for the harvest and then returned to the ports or the mining and manufacturing districts of the Midlands or the north.[106]

Additional tribes of itinerant laborers resulted from the national economy that appeared after 1760, concentrating certain trades in certain geographic regions and forcing laborers to travel from region to region with the seasons of harvest and trade.[107] Harvest laborers generally traveled much of the year, rotating between summer jobs in papermaking in the north and hop picking or harvesting in the south.[108] Failed journeymen turned tinkers adopted the same patterns of sociability as the more organized groups.[109]

Other forms of labor were swept along with the increased waves of mobility among the elite and the military. Carriers and coach drivers fall into this category, as do servants. Audrey Eccles's survey of vagrants in Westmorland found that servants rivaled tinkers in numbers of cases tried, an indication that many domestics worked short jobs and traveled frequently in search of work.[110]

Eighteenth-century attempts to deal with unemployment exposed working-class children to the culture of travel early in their lives. The system of parish apprentices, set up to allocate poor workhouse children to jobs wherever they were available, deployed large numbers of young servants, assistants, and apprentices across the country. Thomas Hollings, a servant born in Coventry around 1765, was leased as a parish apprentice to George Bond of Clapham and was contracted to work for a year. Bond visited a couple of resorts—Bath and Little Hampton—and Hollings went with him as a servant. Hollings's travels with Bond would have exposed him at several points to inn society. The parts of the journey he made with his master would have been spent in the company of carriers and soldiers. He would have walked the first part of the journey from Coventry to Clapham on his own, interfacing with the rest of the nomads who populated that mobile world.[111]

Casual laborers looking for work when employment was scarce enlarged the ranks of other mobile professions that had existed since early modern times: Gypsies and entertainers. Entertainers moved from venue to venue in loose clusters and informal tribes. Most came to the profession between other employments. In the 1760s, Mary Saxby, who ran away from an abusive father and stepmother, lived off discarded cabbages and apples in the market before she was offered the chance to become an entertainer. She became a ballad singer between bouts as a Gypsy dancer and harvest laborer, and she traveled from town to town throughout the southwest performing with another young woman, singing at alehouses, fairs, and markets, until both were assaulted by a group of drunken sailors at Dover.[112] Jem Mace, later a famous boxer, played fiddle at fairs and alehouses as a runaway apprentice in the 1830s or 1840s.[113] The economy of the road was permeable enough to sustain those who were displaced from other home or work situations.

Looser ties still existed among Gypsies, tinkers, and beggars. All three groups were disproportionately watched by the law; in their travels, depending on one another allowed them to escape hardship.[114] William Cameron, a storyteller in Glasgow, traveled with a woman tinker throughout Scotland and Yorkshire in the 1820s and 1830s and earned up to twenty shillings a day helping her solder trinkets. They were asked to join a band of organized robbers on their way to Galway—friends of the woman tinker. Cameron declined, looking forward to a career among the beggars of Yorkshire instead.[115] The boundaries of these outcast nomads were open enough that entertainers and servants could wander in and out of them. Mary Saxby briefly enjoyed a career as a Gypsy dancer when she was seduced by a Gypsy in a fairgrounds; she enjoyed the warm family of the encampment and left only after her husband took a second wife and attempted to persuade her to live in a tent as a threesome.[116] Despite the fluidity of family among outcasts, patterns of mobile dependency extended to beggar, Gypsy, and tinker.

Women constituted a notable community characterized by its exceptional experience. Vivid evidence about the presence of women on the road comes from traveling women preachers who recorded their experiences in diaries. After 1760, Quaker itinerant preachers followed the Methodist pattern of preaching to the unconverted. Esther Tuke of York made fourteen journeys between 1768 and 1789, preaching to large meetings throughout England and into Scotland.[117] A preaching Quaker woman from Hertfordshire in the 1780s traveled a total of 876 miles in seven years, walking between 4 and 31 miles a day.[118] Their narratives recall traveling in company or alone and joyfully meeting other women preachers at the great quarterly events, sometimes traveling with husbands and sometimes leaving the husband in charge of the farm at home for years at a time. The patterns of later women travelers are harder to follow; after 1780, the Methodist Church favored patterns of established rotations by male ministers that meant the extinction of long-distance women preachers, and with them, the major written testimony about women's experience on the roads.[119]

Vulnerable to rape, childbirth, and poverty, women travelers seem to have depended on the ties of community even more than their male counterparts. Outside the safety of a religious community, women travelers traveled with companions united by blood and intimacy. Eccles found female vagrants traveling in families of six to eight, but they also traveled in pairs of close female friends.[120] Mary Saxby describes a world of women harvesters on the road. When she was sick and pregnant one season, she was abandoned by her lover and supported instead by a female fellow harvester, her lover's sister. Female intimacy was apparently more dependable than the casual sexual relationships that developed between women travelers and men.

Outside the better-organized world of mobile communities, the roads were packed with traveling laborers with diverse experiences. From the 1750s, heightened trends of mobility for certain migrant populations, including skilled workers, Scots, and Irish, are demonstrable. The roads prepared the way for new communities whose livelihoods were sustained by travel, including traveling entertainers and casual laborers. The roads also made the wider pool of laborers and harvesters more dependent on the interconnected economy and more liable to seasonal migrations. Through fluid relationships, connected across towns, regions, and networks, poorer travelers of all sorts were part of networks of community less formal and dependable than those of the mobile communities.

Vulnerabilities of Travelers

Travelers entered a realm where they were inherently vulnerable to strangers and unknowns. Merely talking to strangers on the street could have devastating consequences. James Barrett asked a fourteen-year-old girl in Stepney to point him the way to White-Horse-Street, and when they were walking around the turnpike, he threw her and her twelve-year-old sister to the ground behind the hedge at the roadside and raped them both at knifepoint.[121] The archives of petty crimes tried at the Old Bailey are littered with such scenes: the person giving directions to strangers and the stranger traveling alone were constantly under threat of attack by thieves, rogues, and highwaymen.[122] Meetings on the public street offered an occasion for those with ill intentions to mislead, seduce, and betray the innocent pedestrian.

Along with the danger of strangers, travelers also faced dangers associated with towns, where they were often faced with outright hostility and violence. While rotating through the towns, soldiers often alienated townsmen in conflicts over family and sex that have been well documented by scholars.[123] Itinerant artisans faced similar challenges, and Methodist preachers were exempt only because of strict regulations that cautioned preachers against marrying until their career of itinerancy had ended.[124]

Sometimes the differences between soldiers and townspeople were realized simply at an economic level. Innkeepers were rumored to tear down their signs at the report of an approaching regiment.[125] Small wonder: the "Grenadiers' March," sung by carousing soldiers by the 1770s and adopted officially by the Grenadiers in 1815, concludes, "So we'll break mugs and glasses, no reckoning we'll pay,/We'll down with their sign-post, and so march away."[126] This reputation for drunkenness and violence had a simple root: soldiers, likely to pass on to the next town, had different expectations of their relationships than the townspeople who had known one another for generations. Consider what happened when soldiers and townspeople gambled together. Soldiers and townsfolk might gamble, for instance, with less mutual understanding of the rules than two groups of long acquaintance. In Harwich, visiting troops set up hazard tables and enticed several of the townspeople to come and play, "so that in a few Days it became a perfect Gaming house." Trust was broken. A Mr. Collister told the captain that he "lyed like an Irish Scoundrel," and the soldier responded by breaking "his Head with an Oak Sappling." Further mischief was prevented only by his fellow soldiers pulling him away from the other townsmen. Ultimately the accusations had to be settled by magistrates and corporals, and tensions were really relieved only when the company set sail for Flanders.[127]

Officers did what they could to ease such tensions. Officers of men billeted in England or Ireland were supposed to visit the inn where they stayed every payday and "ask the Landlords, if the Men behave well." Other regulations attempted to ensure that the soldiers kept a neat appearance in public by charging the sergeants and corporals with making sure that no soldiers of any rank "appear in the barrack-yard, or street, without their hair being well platted and tucked under their hats; their shoes well blacked, stockings clean, black garters, black stocks, buckles bright, and cloaths in thorough repair."[128]

Such cosmetic regulations did little to disguise the foreignness of discharged troops. As historian Stephen Brumwell relates, "In June 1762, as the Seven Years' War neared its triumphant conclusion, the government-sponsored journal *The Briton* reminded readers of 'those swarms of miserable maimed Highlanders' who could be seen crawling about the outskirts of London 'with scarce any vestige of the human form.'"[129] Many of those discharged from Highland regiments spoke only Gaelic and so were unable to fill out the forms required for their pensions. Scottish poor law was notoriously unable to care for that country's indigent, and so the wandering Scots poor were visited on England.

The differences between soldiers and townspeople were heightened at the close of every war, when large numbers of soldiers, with every expectation of work and support, were thrust on towns that had no means to provide for them. During large-scale deployments at the end of hostilities, masses of soldiers were simultaneously released onto a crippled employment market.[130] At the end of each period of hostilities, the wounded and "worn-out" were the first to lose their place in the army, and with it, their source of income.[131] Many of those recruited and then released came from the most unstable strata of society, to which they returned, wounded and deracinated, less able to find work than before. Those recruited in desperate times tended to be ill prepared, quickly dismissed without pension, and worn out and displaced from their previous livelihood.[132] The ultimate result was a permanent stratum of homeless veterans falling on charity.[133] Of these, many lacked the settlement, or residency, requirement necessary to qualify for poor relief.[134]

Unable to make a living, many turned to crime.[135] Richard Dennison, who had lost his left leg at the battle of Culloden, was indicted for pawning the sheets belonging to his lodging house. He explained that he was driven to do this by the new strictures on loans taken against pensions, which prevented him from drawing money in advance on the sum the army owed him.[136] Wounded and displaced, many soldiers found such petty thefts their only remaining source of hope. The situation crystallized into a pattern of arrests of soldiers in towns at the end of every war.[137] Many others died within the next few years, disabled, wounded, or of unknown causes, living on remuneration as meager as three shillings a day, the lowest rate given to a common laborer.

Methodists were persecuted for a wide variety of reasons and suffered violence of varying forms. In the 1730s, 1740s, and 1750s, Methodists were charged with public disturbance, arrested, and stoned in riots. Such violence most often was explicitly directed at the preachers themselves, men from outside the village, who were interpreted as a vicious and foreign threat. In 1739, John Cennick was assaulted by a butcher who had prepared a vat of blood to throw over the preachers.[138] In the same year, he was later blackened with bruises from a street row. Charles Wesley was stoned by rioters in Sheffield in 1743 who tore down the society house and the house of Charles's host in the town. John Wesley faced riots in Bristol, London, and Wednesbury in the 1740s.[139] In the same year, the justices of the peace at Crowan endorsed the apprehension of the itinerant Thomas Maxfield and some of his tradesman followers for the press-gang; John Wesley himself had to rescue them.[140] One of Whitefield's early converts, the Reverend Henry Tanner of Exeter, was converted in 1743 after he and five or six of his companions heard a zealous preacher in the distance and determined "to go and knock him off from the place on which he stood; and, for the purpose of more effectually injuring *the mad parson*, . . . loaded their pockets with stones." Upon approaching, Tanner was stopped in his tracks by the words he heard and "listened with astonishment."[141] Other preachers were not so lucky. Alexander Mather was nearly killed by a mob in Boston.[142] In the 1780s, Methodist missionaries to Jersey were stoned; in 1787, traveling minister Pierre de Quetteville was told to leave the island by a court trying to punish the attackers. When he asked if he could appeal the decision, he was answered, "We do not allow that to strangers."[143]

Mobile communities tended to disrupt and frighten locals wherever they went, for whenever issues of relationships or money appeared, the separate identity, needs, and expectations of travelers became clear. In the face of these vulnerabilities, community identity became all the more important. Participation in one of the formal networks of the road—Methodist or artisan culture, for instance—cushioned travelers against the worst of these abuses, as did the informal relationships formed between entertainers and women. Confident in being able to trust one another, members of these communities enjoyed the assurance to face new geographies and to travel, knowing that other travelers who identified with them would be there to help. Mobile communities and the laws of identity thus minimized the extent to which most travelers faced rapists, cons, and frauds—by placing those deceits outside the civil expectations of travelers.

Learning Not to Speak to Strangers: The Traveler of the Middling Sort, 1784–1834

After 1784, common travelers were joined in large numbers by travelers of the middling sort, encouraged by the government-appointed system of swift post coaches that filled the highways in that year. Middling travelers moved

"Fruit Shop at Lymington," in Thomas Rowlandson, *Tour in a Post Chaise* (1782).

sheathed within a world of stagecoaches, inns, guidebooks, and novels. Contained within this material zone of safety, middling travelers could restrict their interactions even more by adopting a posture of silent, reserved observation. This posture set the mood for how middling travelers interacted with strangers on the public streets of the towns where they arrived, the only place where middling travelers crossed paths with the deracinated poor. There, middling travelers quickly developed strategies to isolate themselves from interaction with the bodies of the mobile poor: watching them from a distance and paying attention to posture, dress, and gait as clues to where it was safe to interact. Middling travelers thus forged a sphere of safety from the vulnerabilities of the highway by limiting their contact with strangers.

New types of publications, from tourist guides to manuals of etiquette for stagecoach conversation, reflected how late eighteenth-century travelers were cushioning themselves from indiscriminate mingling. The institutions of middling travel, from the coach and inn to the mores and language of observation, were designed to isolate the traveler from the vulnerability of facing strangers. The rise of the mail coach, inn, guidebook, and etiquette book signified a new threshold in how middling travelers were moderating the spaces around them, seeking to establish zones of comfort and sanctuaries of trust among the passing strangers on the modern highway. They developed as an industry around the growing tide of middling travelers after 1784 and were designed to reflect the middling values of speed, separation, and privacy, which depended on enforcing the social difference between moneyed travelers and the pedestrian communities that already occupied the highway.

Thomas Rowlandson's series of drawings, *Tour in a Post Chaise*, documents a trip taken in 1784, the year when the first mail coach was put into circulation by the Post Office, signifying a new era of comfort and promptness of travel that catered explicitly to the needs of the middling traveler.[144] Although the series remained in private hands or lost for most of its life, it exemplifies the road scenes that would typify the rest of Rowlandson's career. In the series, Rowlandson and his fellow artist, Henry Wigstead, travel between London and the Isle of Wight in one of the new post coaches. Interspersed with practical scenes of carriage preparation, loading, and unloading are all the delights that sociable travel had to offer. Wigstead and Rowlandson heckle innkeepers, flirt with barmaids, negotiate with shopkeepers, and race with other carriages. They share newspapers with strangers in a Salisbury coffeehouse and tour the antiquities of ruins and cathedrals. New friends exchange good-byes while waiting to depart from the coaching inn. The artists listen to the innkeeper complain about his business. Wigstead makes new female friends at a fruit shop in Lymington, fondling a lady's waist with one hand while his other hand squeezes a peach. Travelers explored the human and physical landscape within the constricted perimeter of a world defined by consumption.

Improved roads meant that vehicles could travel in every season, not only summer, with relative regularity and speed. Middling travelers were unwilling to subject themselves to the time, dirt, and danger of traveling by foot. Hardy and fit travelers went by horse but were limited by weather and distance. Middle-class travelers followed the coming of new vehicles. It was not until the 1730s that packhorses, the preferred way of getting goods over rough ground since medieval times, were joined by newly designed carriages for carrying passengers over great lengths of territory.[145] The length of coach journeys diminished with improved roads and competition among rival companies.[146] The culmination of these trends was the national, government-run system of mail coaches attempted on a temporary basis in 1784 and widened to national circulation in 1785. Like the stagecoaches with which they competed, mail coaches carried travelers, but exact timetables and reduced schedules quickly established mail coaches as a more reliable and speedy alternative.[147]

Middling travelers sped on lighter wheels at nearly double the pace of slow trade wagons. These vehicles were also the product of government spending and research. As the Post Office expanded in the 1780s and 1790s, the keepers of mail-coach time charts demanded new coaches loaded with springs and more solidly held together.[148] Better-designed, sprung coaches enjoyed enhanced durability and greater passenger comfort and could make better time.[149] Government contracts prompted competition among carriage manufacturers, with the result that travelers marveled at the vehicles on London's streets, which were works of architecture, "with gorgeous panels, superb lining, cushions, &c," astounding the viewer with their display of mobile comfort.[150]

These users consisted of the broadest possible definition of the middling sort: those above poverty who needed to make an important and rare trip to visit relations or tour the town; officers on rotation; sailors on leave from their ships with money to spare; students at Oxford and Cambridge; and judges, businessmen, and lower aristocrats. In a trip Henry Mackenzie took from London, his company "consisted of a grocer and his wife, who were going to pay a visit to some of their country friends; a young officer who took this way of marching to quarters; a middle-aged gentlewoman, who had been hired as housekeeper to some family in the country; and an elderly well-looking man, with a remarkable old-fashioned periwig."[151] Robert Southey traveled with "a member of the university" and "a fat vulgar woman who had stored herself with cakes, oranges, and cordials."[152]

For middling travelers, speed was becoming a commodity to be bought with cash and protected by law. Speed and design heightened the differences among vehicles of different classes. Coasting atop a mail coach "at so delightful a rate," with "nothing to do . . . but to enjoy" himself, one traveler pitied "every foot passenger" he met.[153] The vehicles of the middling sort in London moved at twice the speed of the floods of laboring "carts continually passing with goods from the different wharfs on the south side of the river Thames." The stream of vehicle traffic was incredibly diverse, including "cabs, tax-carts, gigs, trucks, drays, carts, wagons, omnibuses," and donkey carts, "top-heavy with baskets," ferrying milk, ale, coal, carcasses, and vegetables to various commercial destinations.[154] When the question of establishing a speed limit on the carriages of the elite was raised in Parliament in the 1830s, it was condemned as an absurdity, "a flight in legislation," an unnatural attempt to make the swift and the slow go at the same speed, a law impossible to enforce, "as no sane person ever before dreamt of."[155] The slow were wise to hug the side of the road, moving in slow packs, avoiding the swift, whose difference was presumed to be natural. The fast middling sort, contemporaries understood, was a separate breed from the slow common traveler, and the two classes existed in a natural rivalry or predatory relationship.

Those who mounted stagecoaches left the dirty world of the street as they ascended into the secure architecture of the carriage's moving world. Robert Southey felt safer from the "rogues of every description" who swarmed around the street as soon as his luggage was safely stowed in the boot under the coachman's feet and he himself was seated in the stagecoach.[156]

The swift vehicles of the middling sort and the aristocracy were a danger to pedestrians. Despite the footpaths raised on a curb regulated by Parliament, horses drawing vehicle traffic came so close to pedestrians "that their hoofs, and the great wheels of the wagons," were "only a few inches from" the people walking on the footpath.[157] The flaneur James Grant admitted that this speed was a terror to the ordinary pedestrian of any class: "The stranger fancies every moment that some one will be run over, or that some serious accident will take place from their coming in collision."[158] Sometimes, indeed, speed had violent and terrible consequences for pedestrians. William Hoe of Brick Lane testified in 1840 that he had been walking along the raised footpath above the roadway in Highbury with his wife and five-year-old daughter Esther one bright evening when a racing hackney cab crushed the child beneath its wheels.[159]

Casual travel, having entered the realm of the middling traveler after 1784, quickly was perceived as a zone of social difference where speed distinguished swift travelers from slow. It also became the site of an uneasy tension where middling travelers learned to negotiate their space from their fellow passengers. The social rituals of speed and small talk cushioned middling travelers in an alien world entirely distant from that of the mobile poor.

When the coach stopped, all alighted at the inn, where they could sup with the same people with whom they had come or discourse with the innkeeper and his family. Middling travelers, letters, and goods were deployed and received at a network of inns. Inns served as the hubs of the road network, where travelers rented horses, changed coaches, and stopped for rest and refreshment.[160] Seventeenth-century inns had brought travelers directly in contact with local cultures of sociability and politics. Aside from their function as post offices, inns functioned as the major venue for proto-music-hall concerts, local skittles and quoits competitions, and billiard playing.[161] But the traditional inn, with its diverse roles as a center for village and town, was quickly disappearing. Eighteenth-century inns began catering to a set of more diverse travelers. In 1746, Paul Sandby, traveling to Scotland as a draftsman, stayed at inns when he stopped at various places to make sketches of the landscape, which were later collected by Joseph Banks.[162] The draftsmen John Hassell and Julius Caesar Ibbetson traveled together through the inns of London and its boroughs as they sketched the area's canals and ports.[163] These travelers—essentially part of the auxiliary of military soldiers who frequented the same inns used for billeting—were soon joined by occasional business travelers, diplomats, and naturalists, who recorded their travels and conducted a lively community of exchange centered on the inns.[164] Such travelers, who often spent long sojourns at the inns that they made their base of operations, formed a community often more cosmopolitan than that of the towns through which they passed.[165] Exposed to a broad community of travelers, the sons of innkeepers became naturalists, scientists, landscape painters, actors, and travel writers in unusually high numbers.[166]

Mail-coach travel gave rise to a new variety of custom-built hotels catering solely to the needs of business travelers. Glasgow's Tontine Inn was founded in 1781 by subscription by a group of English commercial travelers and bankers who wanted their own society and thereafter used the inn as a clubhouse and general exchange as well as a hotel. Describing the elegant classical reading room, one visitor was shocked by its difference from the homely taverns provided for travelers in other cities: "Continually filled with readers and persons who frequent the place on business, it exhibits a spectacle as surprising as it is amusing."[167] After 1790, the Star Inn in London, near the docks, became a major center for businessmen dealing with the trade in the ports.[168] The Angel Inn, Islington, was the particular resort of "the Salesmen, Farmers, and Graziers attending Smithfield Market."[169] Another traveler found the Talbot Inn in Water Street, Liverpool, "adapted to the man of business" but "rather too noisy for a studious man."[170] The sum of these changes pushed inns to change their appearance, putting up classical fascias designed by classical architects, separating a variety of reading and dining chambers, and ensuring the privacy of travelers. In short, the movement from inn to hotel was complete by 1820.[171]

Even in their architecture, inns offered a reassuringly secure experience for the middling traveler. Established as a courtyard in Roman times and fixed in an English idiom of storied galleries in the Middle Ages, the same architectural format was lent to the hundreds of inns in brick and stone that appeared throughout the country during the eighteenth century. Holding to the traditional form allowed the inn to be distinguished from surrounding houses. Flamboyant inn signs and arches across the road signaled turnoffs to fast coaches, first appearing piecemeal in a few locations and later forming a replicable pattern—like the overwrought hunting-scene arch over the road at the White Hart, Scole, Norfolk, of 1655—generalized by the mid-eighteenth century into a familiar cross-road iron sign.[172]

"When we wish to travel," wrote a reader of the New Monthly Magazine, "'Maxims of Locomotion,' 'Guides to the Watering-places,' and 'Companions' to every county in England . . . contend for the honour of directing our steps."[173] Frederick Reynolds, the comedian, explained in his autobiography that all middle-class journeys began with "those three grand preparatory delights, packing up, making purchases, and consulting road books," a ritual of approaching the trunk makers around St. Martin's-le-Grand in London and buying one of the well-distributed guides like Daniel Paterson's, which went into new editions every year after 1790.[174] They became indispensable to travelers who were visiting new areas for the first time.[175] Reynolds joked gloomily that the preparations of book buying and map reading remained "probably, after all, the traveller's best amusements."[176]

Maps were a tool that distanced the traveler from dependence on hired guides and helpful innkeepers. In the days before the coming of accurate maps, travelers learned about the route ahead from their host at the inn.[177] The first travel-sized road books gave limited information about the roads as a network. Sometimes luxury items painted on expensive vellum sheets, road books catered to the elite.[178] From the 1670s, small foldout maps appeared in the pocket guides carried by travelers. These miniature volumes offered the image of a kingdom joined by a road network.[179] Their audience was small, well traveled, and polite. They needed less topographical information and probably relied on hired guides to help them navigate between towns. Poorly corresponding to actual roads, the pocket guides supplied a vague notion of how to plan an itinerary but offered little practical help to way finding on the road. The early eighteenth-century road books focused on itineraries—the path from one town to the next—rather than on the road network as a whole. Users of Emmanuel Bowen's *Britannia Depicta* (1764) would have little sense of how crossroads they passed in their route corresponded to the shape of a larger network and a national whole. Strip maps were rather meant to enrich the path of a single possible route, perhaps already familiar to the traveler. Long passages in the margins of John Mostyn Armstrong's strip maps detail the historical antiquities of the area and anecdotes about gentry houses. Early pocket maps thus combined limited information about topography and nation with scattered information meant to satisfy travelers' picturesque, historical, and political curiosity. They still presumed a culture of conversation where necessary information about the shape of the road network as a whole, the culture of hostlers and innkeepers, the expected rates of payment, and the times of the journey would be supplemented by talks among travelers on their way.

Maps of the road network as a whole, which allowed travelers to plan their own itineraries, signaled a new kind of journey where travelers expected to choose their routes rather than asking for help from innkeepers or guides.[180] Beginning in the 1780s, larger, folded maps of the country could show the road network with enough detail to simultaneously depict the

John Ogilby, *Ogilby's and Morgan's Pocket-Book of the Roads* (London: Printed for J. Brotherton, 1752).

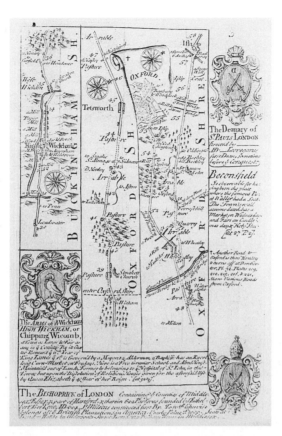

Road to Oxford in John Ogilby, *Ogilby's and Morgan's Pocket-Book of the Roads* (London: Printed for J. Brotherton, 1752).

roads as a network joining long distances, provide detailed accounts of distances, and allow comparisons among different towns. The more widespread use of network maps followed exactly the spread of travel and post. In 1784, the year of the first mail coaches, the printing industry began to produce a new kind of map, graphically sophisticated, nationally oriented, and cheaply produced for wide distribution. Daniel Paterson, John Cary, and Carington Bowles, three London printers, were to dominate this new scene of printing with a flood of cheaply printed maps, characterized by the single-sheet foldout road map of the whole of England.[181] By 1798, wall-size maps of England in these convenient forms were on sale for three to five pounds in London. Reduced-sized images sold more reasonably for two shillings.[182] Reprinted every year between 1771 and 1832, Paterson's books claimed an accuracy, official circulation, and cheap reproduction that set them apart from the map books of the elite.

The new maps differed by their precise calculations of distance, their attention to topography, and their representation of the road network as a whole. Only this scale of detail could serve the needs of carriers, merchants, and travelers, who needed to compare different possible paths—with relative distance and directness—at a single glance. These were not enriching diversions for travelers already familiar with the route, but tools for travelers who were freed from dependence on others.

At roughly the same time, the antiquarian functions of strip maps combined with the function of travelers' diaries to produce a new entity: the guidebook. Cary's and Paterson's maps were topographical tools stripped

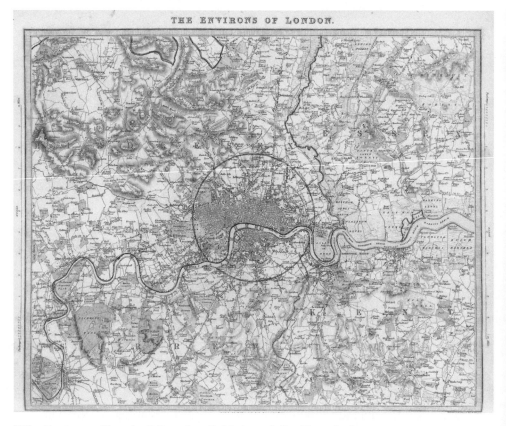

"The Environs of London" (London: Baldwin and Craddoc, 1832).

of the elaborately calligraphed passages of local history in the margins of old maps. The new guidebooks anthologized botanical, historical, antiquarian, and geologic information, synthesizing the different kinds of facts that one had previously received by inquiry in the diverse social concourse of the inn. Ordinary travelers could therefore supplement information from their fellow travelers with the company of expert ones. In 1773, Thomas Pennant published his popular *Travels*, the record of his expeditions in the company of a Gaelic expert, a botanist, and a draughtsman; Pennant himself provided the political and historical commentary. Painter-diarists like Joseph Farington illustrated the tours as Pennant's work went into multiple editions and was emulated by followers.[183] By the 1780s, such guides had diversified into spiritual, philosophical, and literary varieties as well, supplemented by an additional fleet of books promising an accurate account of weather, geology, and botany for the itinerary in question.[184] Robert Southey amused himself by observing "the frequent towns, the number of houses by the road side, and the apparent comfort and cleanliness of all, the travellers whom we met, and the gentlemen's seats, as they are called, in sight," and found these sights enriched by his guide book, which elaborated "every one" of them.[185]

Combining the voices of the most articulate among government, business, and leisure travelers, such guides rendered inn society in the convenient form of the book. The guidebook abstracted information that originally had to be gained from human sources. With guidebooks and maps, middling travelers could interpret the route and landscape without having to resort to social intercourse.

The Rule of Etiquette

Outside professional business, meetings in inns were kept haphazard and brief by the schedule the coaches kept. John Gamble found the Britannia Arms, the great inn of Shrewsbury, uncomfortably "crowded with passengers of the Holyhead Mail." He recalled, "To a delicate person their clamour might have been annoying, but the annoyance of a mail coach company can never be long; the horn sounds, and, like ghosts on the crowing of the cock, refreshments scarcely tasted, they must hie away."[186] The strict schedule kept by post coaches meant that travelers rarely had time to interact with locals or the village. "The rapidity with which our stage-coaches now travel has almost driven away all conviviality on the road," explained one traveler, "for should hunger drive you to dine, you are forced to devour your victuals like a cannibal, and then run like a debtor pursued by bailiffs. Laughable incidents frequently occur from the shortness of the time generally allowed for refreshment."[187] A pattern of quick movements therefore typified the visit to an inn. In A *Visit to Edinburgh*, the stagecoach travelers stopped at an inn for a change of horses and were instantly conveyed "into a neat small parlour, where they dined whilst the chaise was preparing, into which they hastened as soon as it was ready, and were quickly seated as formerly."[188] These rushed schedules meant that post-coach travelers had little experience of the villages that intervened on their itinerary.

Stagecoach schedules made for arbitrary times, late arrivals, and early departures that left travelers in little state to interact with the world around them. Robert Heron joked, "It shakes one into habits of subordination and obedience, by subjecting you, for the journey, to the command, in fact, of the mail-coach driver and his horses."[189] The grueling trips from Scotland, driving through the entire night, were particularly bad, delivering travelers to London inns in the wee hours of the morning and leaving them exhausted and bewildered as they explored the city. After two days and three nights in a stagecoach from Edinburgh, Thomas Staunton St. Clair "arrived at six o'clock in the morning at the Post Office, in Lombard Street," and tumbled "immediately into bed at the Bull and Mouth." For St. Clair, "The whole of that day was spent in a sound sleep; nor did I once open my eyes again until nine in the evening, when I found it too late to execute my commissions."[190] He recalled, "I immediately went to bed, breakfasted early in the afternoon, and took a long country walk."[191] After a long ride from Perth, George Cooke arrived at the Bull and Mouth "about five in the morning of Tuesday." Such schedules winnowed all but a few impressions of the inn staff who rushed to cater to the arrivals' minimal needs—the way the "landlord and landlady make their appearance with smiles" as the coach "rattles" "under the gateway into the inn-yard . . . and the waiter turns round the brass handle of the chaise-door in haste, to hand you down the steps."[192] Napping away the wan hours between arbitrary arrival and departure times, travelers had little occasion to meet one another on sociable terms.

Middling travelers in fact enjoyed one another's company only in the stagecoach. Stagecoaches mixed the full range of middling travelers, from housekeepers to judges. Such mingling in cramped quarters over the course of hours could give rise to physical discomfort. The passengers seated outside on top of the coach were liable to be poorer and noisier than those seated within. A traveler on the Liverpool coach to London joined "a number of seamen, who drank, sung, and quarreled during the whole of the journey; I do not suppose there ever was a more noisy coach, since coaches were first invented; a mill was the temple of silence in comparison."[193] But conditions inside had their own peculiar discomforts. One traveler found himself crammed between "three fat, fusty, old men," "a young mother and a sick child," "a cross old maid," "a poll-parrot," "a bag of red herrings," and "a snarling lap-dog."[194]

Meetings in stagecoaches rarely inaugurated longer relationships. Even in the best cases, those who had enjoyed pleasant conversations entertained only "a few slight regrets . . . at the moment of separation," explained one magazine editor. More often, he wrote, "We look with indifference, and sometimes with cold neglect, at the loss of our acquaintance, and often do not suffer even a sigh to escape us." He summarized: "A stage-coach may very properly be compared to the world at large;—we breakfast, dine, and sup together, a few times at most, and then part, to meet no more."[195] Indeed, manuals of etiquette warned that "meeting and conversing with a person in a stage-coach . . . does not entitle one to salute him when he is casually met afterwards; and it is a mark of ill-breeding to claim acquaintance on such grounds."[196]

In the 1810s, these interactions began to inspire encyclopedic catalogs of the dangers of intermingling on the stagecoach. Memoirists and comedic writers stuffed their volumes with anecdotes stressing the violence done to them by strangers.[197] "It instructs us, not to be squeamishly over delicate in favour of sweet odours," joked Robert Heron. "Crammed up in a mail-coach, with males and females of every diversity of years, health, and condition, we necessarily learn fortitude, in the exercise of all our senses, but especially the sense of smell."[198] So many people, children, and pets made for a potential sensory overload of a highly unpleasant nature. Uncouth children added another hazard. One traveler recalled how, as a twelve-year-old boy, he had intentionally taken advantage of stagecoach journeys to experiment with sadism, "slily" dropping "the gravelly dirt I had collected from my shoes, down the neck and back of a very pretty girl, who sat blushing furiously on my left."[199] Such stories appeared in years of heightened political tensions and typically emphasized the friction caused by having to deal with the diverse public opinions of the stagecoach's other passengers. Robert Southey related, in the persona of Don Espinosa, how his fellow passenger exhausted him with her naïve views of the war, French brandy, "Bonniprat," and whether papists read the Bible.[200] Sardonic stage-coach stories pointed to how political diversity pushed the limits of sociable intercourse that had typified traditional travel narratives.

By the 1830s, etiquette writers began to offer advice on how to filter the pattern of delimited conversation established in the stagecoaches.[201] One such writer, explaining at great length the gallantry with which Englishmen gave up their places to women, advised young ladies to inquire, where possible, what they could do to better convenience their fellow passengers.[202] One could attempt to avoid the worst of such interactions by carefully observing one's fellow travelers for clues about their identity, but even looking too closely could be construed as an insult. Such patterns of observation had been developing since the stagecoach rides of the late seventeenth century, and Joseph Addison remembered in the *Spectator* how he amused himself when he had "no inclination to talk much" by sitting across from his fellow passengers "with a design to survey them, and pick a speculation out of my two companions."[203] Nineteenth-century travelers had even more tools with which to make such observations. An etiquette writer advised the careful watching of "inn-keepers, stage-coach proprietors, drivers, waiters, chamber-maids, and fellow-passengers" in order "that we should have clear ideas of the rights of those, whose interests are most likely to clash with ours."[204] The budding science of physiognomy could also theoretically be applied to such problems of interpersonal negotiation. Henry Mackenzie introduced his friend Harley, who, whenever seated in a stagecoach, would "set himself to examine . . . the countenances of his companions." This "inclination to physiognomy" tended to alienate him from other stagecoach passengers, who were piqued by his act of staring in critique.[205]

"Small talk" became the mark of the well-adjusted middling traveler, a sign that one could perform the gentle dance of conversation without giving or risking offense or monopoly. A traveling Oxonian might therefore entertain himself by "amusing the ladies with a variety of small talk."[206] It might consist in describing the stories of contemporary business successes or the makes of various carriages.[207] It fixed "upon the state of the weather, and of wind," and although it scandalized aristocratic literati who thought it the height of dullness, it was an essential technology for the survival of those forced to travel long distances.[208] Middling-sort etiquette books therefore defined "small talk" as a characteristic of "good company," marked by "a certain manner, phraseology, and general conversation."[209] It stood out as distinct from "civility" and "formality," neither of which courtly arts aided the traveler who was attempting to minimize the potential differences of the diverse company around him. A journalist explained its application to mediating "mixed society, where men and women, young and old, wise and foolish, are all mingled together."[210] Small talk ideally aimed at pieces of knowledge to which everyone had equal access. It therefore failed when one added any information of expert or professional knowledge; the same Oxonian in the coach made the mistake of referencing Sophocles amid his conversation and ended up trapped in a deep and endless monologue on Aeschylus and Euripides by the Cambridge don seated across from him.[211] Small talk was a delicate business that was successful only when it moderated such interminable stretches of arbitrary proximity.

Small talk evolved in the new environments of mobility—"in stage-coaches, at hotels, and at watering-places"—wherever middling travelers tried to control their circumstances through the technologies of conversation.[212] As another etiquette writer explained, "In no country in the world is the proportion of travellers to the population so great as in ours; and therefore it is peculiarly incumbent upon us to understand the morals of traveling," and therefore the art of conversation.[213]

Travelers of the middling sort could buy their way into safety from the worst deceits of the road, relying on inns and coaches to keep them from outright deception and murder. Printed tools like maps and guidebooks gradually replaced the traveler's reliance on human conversation. This cycle of isolation was finally completed by a shift in etiquette. Within the protecting bubble of the stagecoach, travelers were still vulnerable to bores and zealots. Aided by observation and guarded by small talk, middling travelers began to identify safety with isolation from interaction. As their numbers increased after 1784, middling travelers developed layers of institutions that cushioned them from the uncertainties, frauds, and bores of the road. The culmination of those efforts was not to speak to strangers except in the tentative exchanges of small talk: the middling traveler had learned long since that watching them in critical silence was safer.

Making Travel Respectable, 1822–1848

As the highways became the domain of middling travelers, a series of regulations gradually extended the powers of local police over the public spaces of the highway. By the 1790s, even veterans, long exempt from prohibitions against begging, could be imprisoned, whipped, or deported if they were caught panhandling in public spaces. By the 1810s, being suspected of unemployment away from one's parish of origin was crime enough for members of the mobile poor lingering around docks or public streets. By 1820, English parishes had criminalized mobility for the poor, initiated an era of feral accounting of comers and goers, and begun to collaborate across geographies to trade information with other urban managers in relation to control of the mobile poor. The 1822 Turnpike Act imposed fines for camping at the side of the turnpike (this provision was extended to all highways in 1835), thus criminalizing the mobile behavior of most tinkers. The Vagrancy Act of 1824 outlawed "sleeping rough"—that is, sleeping out-of-doors in barns and fields, as homeless people do. The Vagrancy Act thus rendering criminal the entire class of poor travelers as a whole. New acts extended the power of the police to punish members of the mobile poor caught sleeping rough or suspected of criminal activity. The Vagrancy Act extended the conditions upon which a highway parish could send poor travelers to their native parish. Under the act, the working poor could no longer gain residency upon being hired for work. By abolishing settlement by hiring and service, its provisions required out-of-work laborers to move on instantly to where work could be found, lest they be forcibly deported to their original settlement.[214] The act thus required poor travelers who used the road to use it only as a conduit from job to job, not as a site for begging, loitering, or depending on their own community. Any member of the poor who fell in with the mobile poor and thus ended up on poor relief far from his county of origin was criminalized by the system. Remarkably, punishment fell on all heads equally: not only Scots or Irish, but also poor veterans, idle Methodists, Gypsies, harvesters, ballad singers, and entertainers; the entire lot of the traveling poor was classed as dangerous. By 1810, any traveling member of the poor, employed or not, was viewed as a potential problem. Only once the road was cleansed of undesirables could it be represented and experienced as respectable.

Indeed, the politics of road building had already produced a movement to envision how the road could serve as a producer of civility. Advocates of centralized building suggested that they could trace the relationship between transport and moral progress. James Anderson, a correspondent of John Sinclair's, insisted in 1790 that he could discern an exact relationship between roads and well-being: "Diminish the expense of carriage but one farthing, and . . . you form as it were a new creation, not only of stones and earth, and trees and plants, but of men also, and what is more, of industry and happiness."[215] In 1831, as further measures for centralization were being mooted in Parliament, the *Metropolitan* reckoned that as a result of roads in Scotland, "the country has, within the course of a few years, been advanced above a century in improvement and civilisation." The author explained, "The moral habits of the great mass of the working people have been changed; they have been taught to depend upon their own exertions for their own support, which is among the greatest blessings that can be conferred on any people."[216] A witness for a parliamentary committee on Irish roads testified in 1835 that when he had first seen Highlanders, in Sutherland, they were mostly poachers and smugglers, but after getting roads, "they have given up both, and have become most industrious workmen in every class of agricultural labour."[217] He bore witness to "new houses of a better class" and "the greater extension of the English language." The author of *The Roads and Railroads* (1839) declared that his object was "to show that the improvement of mankind, and the perfection of the means of internal communication, have progressed simultaneously."[218] A road propagandist described the "large crops of wheat" that now covered "former wastes" and explained that roads in Scotland had "done more for the civilization of the Highlands than the preceding efforts of all the British monarchs."[219] William and Robert Chambers claimed to have discerned that roads were "the most powerful of all agents of civilization and social advancement" and that the more they were improved, "the more rapid and effectual will be the spread and increase of social and intellectual happiness."[220] George Richardson Porter, in his 1836 work *The Progress of the Nation*, boasted that roads had resulted in "a great improvement" related to "the greater extension of the English language."[221] Even the working class would be included in this assimilation project. Pathways as neat as any "garden path" were now brought down to the level of common highways, "used and enjoyed, not only by royalty and nobility, but even by the humblest of our race."[222] Transport, economic progress, and moral assimilation went hand in hand.

For believers in this version of modernity, confident assertions about the effects of roads bolstered a vision of the nation's present identity as a people unified by the middle-class values of speed, engineering, and progress. Porter boasted, "There is not any circumstance connected with the internal condition of England which more strongly excites the administration and the envy of foreigners than the degree of perfection to which we have brought our means of internal communication."[223] "Let any one watch the immense number of stage coaches which, starting at one common centre, London, fly in all directions through the kingdom, making distances, however lengthened, disappear in the rapidity of their movements—and he will immediately perceive that the facility of communication has in fact brought the towns of the most distant counties to be no more than the suburbs of the great capital."[224] Explaining the effect of "transit and intercourse," Michael Angelo Garvey could define his Britain in terms of a successful project of unification and assimilation that had destroyed the ancient hatreds among counties, regions, and former colonies: "The nation becomes an organic unity, the vital energies circulate from the great centres of social life to its remotest extremities, and all its movements are consentaneous."[225] Unlike in the old world of superstition, where men repeated notions told them by strangers, modern and civilized Britons could identify truth in the bodies of their official experts. By the 1850s, these conjectural histories had developed into a full-fledged mythology of British civilization. Standing back "like the necromancer" to "call up the spirits of old to speak for themselves," the editor of *Blackwood's* could discern that "the art of road-making, and postal regulation, and the establishment of mail coaches, were the most satisfactory means of advancing civilisation."[226] Describing the progress of roads from "tracks made at random through the fields" to the "utmost perfection under McAdam," Garvey argued "that communication had visible effects . . . in the manners and customs of the people, in their social and moral relations, and even in the laws of the land." He defined the coming of roads against an ancient past, "when notions and superstitions were inherited with the soil, and descended without interruption from generation to generation, unchanged by contact with strangers, and untested by their investigation; when men generally lived their whole span and performed all their life's labours within the compass of a few miles, and rarely knew anything of what took place beyond the boundaries of their native parish or county."[227] Telling such a story about how roads had transformed Britain, historians could assert that they lived in a nation integrated by consensus about the middle-class experience of travel, trade, and moral progress.

It was Thomas Babington Macaulay who took these arguments to their furthest possible extent. In the famous third chapter of his *History of*

England (1848), he set forth the shape of an argument that would be rehearsed uncritically for a century to come: that since 1688, Britain had been engaged in a process of consensual enlightenment that had been set into play by the roads, mobility, and the circulation of print. These forces, Macaulay argued, had transformed Britain into a sociable nation united in spirit, and in 1848 this unity of expectation was manifest in the nation's freedom from those violent revolutions seen across the Continent.

Macaulay wove this story by filling in vivid details about how much the roads had improved circulation, and how this mobility implicitly resulted in sociable intercourse and understanding. For the rise out of poverty, corruption, and despair to represent a complete revolution at the hands of the engineers, Macaulay had to exaggerate the horrors of the pre-highway past. Circulation had been virtually nonexistent. The smallest trip involved "such a series of perils and disasters as might suffice for a journey to the Frozen Ocean or to the Desert of Sahara," and "it happened, almost every day, that coaches stuck fast, until a team of cattle could be procured from some neighbouring farm, to tug them out of the slough." Macaulay suggested changes of magnitude by implication rather than by fact. The sticking coaches and deserts of the imaginary past were described in detail; the contemporary improvement was left vague. The historian preached that the seventeenth-century "mode of travelling" on coaches sticking in the mud "seemed to our ancestors wonderfully and indeed alarmingly rapid," quoting seventeenth-century advertisements for the first flying coaches. In contrast, contemporary Englishmen placed on the same vehicles would regard them as "insufferably slow." Everything had been improvement, he argued, castigating the past. So too, Macaulay alluded to an even more important shift of morals, away from "stupidity and obstinancy," when travel was imperiled by danger and highwaymen, toward a new age of modern travel characterized by "luxury," "liberty," and "jollity." By implication, then, contemporary Great Britain in 1848 was as sociable, rich, and safe as any nation on earth, and certainly more sociable than it had been at any era in its past. The lower classes as well, Macaulay asserted, had been miserable in 1688. All these tensions had been relieved, Macaulay was convinced, by the influence of trade and transport. The "mollifying influence of civilization" had made people more sociable than they had ever been in the past.[228]

A master of rhetoric, Macaulay knew how to manufacture a story. The grand sweep of the road revolution that he offered—a revolution in mind, manners, and democracy—necessitated kinds of social evidence that did not exist: evidence that Britons talked to one another more in 1848 than in 1688, when in fact the opposite was true. To make up evidence where he had little, Macaulay resorted to the theatrical device of putting himself in the seventeenth century, fictionalizing an account of the London of that age, describing it as a city of the lost, where interaction was characterized by trauma, brutality, violence, and loneliness.

Macaulay stressed the exoticism of his ancient forebear, employing hyperbole to highlight the isolation of the past: "A cockney, in a rural village, was stared at as much as if he had intruded into a Kraal of Hottentots." Even a lord from as far away as Lincolnshire or Shropshire was comparatively as strange on the Strand as "a Turk or a Lascar." Macaulay fabricated the visual and material detail that would have distinguished such seventeenth-century provincials, pointing to "his dress, his gait, his accent, the manner in which he gazed at the shops" and "stumbled into the gutters."[229] The historian was raking his memory for all the visual details that characterized exotics on the public street and was coming up with posture, gait, and dress, characteristics that London's writers and illustrators had learned to distinguish only two generations ago.

Macaulay then turned to the violence he imagined a traveler would face. Because the body in the city was legible, Macaulay understood, it was vulnerable. The traveler's exoticism "marked him out as an excellent subject for the operations of swindlers and barterers" and subjected him to exploitation by thieves and lowlifes.[230]

Moreover, the tendency for strangers to read the traveler and exploit him degenerated into a culture of hostility and isolation, characterized by the disintegration of trust. Macaulay imagined a city where strangers spoke to one another at great risk. If a traveler "asked his way to Saint James's, his informants sent him to Mile End." Shopkeepers tried to persuade him to purchase "everything that nobody else would buy," down to "watches that would not go." Even sites like "coffee houses"—where he should have been greeted—offered the stranger hostility in the form of the "derision" and "waggery" he faced in conversation with those more urbane than himself. Such a city, Macaulay warned, encouraged a civilization that preferred isolation and prejudice to exchange and civilization: "Enraged and mortified, he soon returned to his mansion, and there, in the homage of his tenants and the conversation of his boon companions, found consolation for the vexations and humiliations which he had undergone."[231] The city of the lost was a public madhouse where police, crooks, and bullies judged and threatened the body until, in flight and isolation, the traveler artificially enforced his privacy by returning to his own house.

It is jarring to read this sketch after being immersed in descriptions of the London in which Macaulay grew up. To paint the city of the lost that Macaulay claimed was seventeenth-century London, the historian transcribed descriptions of contemporary London. Even the language he used, of "jostling" and "stumbling," consisted of recent slang words borrowed for the precise identification of different ways of moving through the city. The characters who threatened the traveler—"bullies," "thieves," "moneydroppers," and "painted women"—were the raw synthesis of a century of city canting books and instructions for travelers, most of them more modern than ancient. The ultimate source for these fictional details was in fact not the distant past, but Macaulay's own experience on the nineteenth-century roads.

The *History of England* asserted that highway travel was a linear movement in time toward civilization and sociability. Its success as a work of history rendered it an icon of belief, and so a force for change, as middle-class Britons came to believe that they indeed inhabited a sociable nation unified, rather than divided, by its roads.

CONCLUSION

The Necessity for Infrastructure

> We are full of empires within empires: they may have some good effects, but they do not make good roads.
> —Review of CHARLES DUPIN, *The Commercial Power of Great Britain, Westminster Review,* October 1825

For better or worse, in the modern era, infrastructure unites distant strangers into new communities. Following Britain's model, nineteenth-century nations recognized the necessity of connecting their peripheries, and centralized railroads and government land grants ensured the production of similar connective networks across Europe, North America, and their colonies. But infrastructure also divides the nations it nets together. As economists have reflected on the problem of the "public good," the highway has frequently appeared to refine the question. In the eyes of many, the benefits of highways to the public were more impressive than their threat. To historian of technology Michael Angelo Garvey, writing in 1852, the age of highways foretold an era of communications in which all languages and nationalities would eventually collapse into one.[1] Others considered the role of interest in massive works of infrastructure too important to ignore. J. S. Mill, writing in 1859, saw highways as clear manifestations of the modern expansion of bureaucracy, indissociable from the "great evil of adding unnecessarily" to the power of the state over the individual.[2]

Such differences in perception date from the beginning of the infrastructure state and reflect geographic and economic differences in the needs of citizens. Road partisans tend to benefit directly from the creation of infrastructure. In eighteenth- and nineteenth-century Britain, highways appeared to lobbyists like Henry Parnell and John Sinclair as the means by which states could unify warring communities and simultaneously ensure their perpetual economic growth. Adam Smith argued that by breaking down the barriers to economic participation, government infrastructure would encourage national security and prosperity. It was a vision so compelling as to make state-built infrastructure a necessity for all modern nations, a defining feature of the experience of what it means to be modern.

But not everyone benefited as equally as claimed from this vision. In nineteenth-century Britain, libertarians and radicals raised important questions about who would govern this new commons and how. Libertarians objected to the way expert rule displaced decision making by political consensus. Infrastructure remained vulnerable to the greed of self-interested elites who wanted the benefits of connectivity only for themselves. Meanwhile, radicals objected to the cost of expert-designed infrastructure, arguing that the poor could little afford to pay the cost of a new toll. Geographic participation of the periphery in the market was at odds with the economic participation of the poor. Every nation swings between these poles. Far from unifying the nation or ensuring economic destiny, the era of infrastructure is marked by contestation of the state and the radical reversal of economic fortunes.

Clamoring for parish representation and the free market, radicals like William Cobbett and libertarians like Joshua Toulmin Smith challenged Britain's budding bureaucracy. Their powers of persuasion dismantled the infrastructure state only some forty years after it was first imagined. Between 1836 and 1880, no new roads were built at parliamentary expense. Between 1836 and 1864, the libertarians, urged on by Joshua Toulmin Smith, won the day. Their outlook came to dominate every facet of English politics for the second half of the nineteenth century, organizing the principle of local government in public health, the poor law, and the police.[3] By the 1860s, road building was entirely a local responsibility, and the few centralizers left failed in their feeble attempts to channel supplementary funds to the remote parishes that struggled to maintain their roads.[4] The home of the Industrial Revolution became a study in contrasts. Rich cities like London were woven with tunnels and viaducts, while the streets of poor towns on the nation's outskirts became bogs at the first sign of rain. Localism had won the day.

Localism caused the abandonment of roads, destroyed the Scottish economy, and left broken dirt trails in the very farmlands where modern

CONCLUSION

infrastructure had been invented. The swing away from the state was marked by a degree of naïve conviction. As Britain dismantled the infrastructure state, contemporary historians looked back with certainty that the medieval, Tudor, and modern parish government had produced a realm of participation, justice, and freedom rather than exclusion.[5] These gains came at the cost of pricey tolls and failing roads, and libertarian victories were a rout for rural communities and the poor.

Eventually their discontent led to a second reversal. Driven by the high cost of toll roads and the near impossibility of earning a living, poor Scottish farmers organized the massive road reform movement of the 1880s, reappointing the dismantled infrastructure state. From centralization to localism and back again: so swings the pendulum of attitudes toward the state. Such reversals typify the fate of the infrastructure state. The tensions between connectivity and economy, between expert rule and local control, govern the history of infrastructure.

The wide-eyed apostles of libertarianism pushed to their limit claims about the virtues of decentralized roads. Free-market theorists proposed reasons that local governments, left to their own devices, would abolish the expensive tolls that plagued the poor. The theory was that landlords and shopkeepers would object to high tolls or poor roads as soon as they adversely affected trade.[6] The expectations were straightforward. There would be cheap roads where people traveled, paid off by reasonable tolls, and no roads where the people needed none.

Free-market advocates even promised that the minimal legislation would chase the hated payments of tolls out of existence. Turnpikes would collect tolls only until their lease was up. At that point, the leaseholders would be paid off, and the road would become parish property again, efficiently supported by minimal taxes. Soon, cheap and efficiently built roads, controlled at the local level, would freely connect the poor of every vicinity.

All that was required, according to free-market theory, was minimal government oversight to ensure that turnpike trusts reported their takings. The argument persuaded many listeners. Select committees in 1836 and 1864 agreed that local magistrates, elected by local ratepayers, would build roads for "the most advantage to the public."[7] Localist legislation was passed in 1851 and 1852 that focused on requiring turnpikes to report their progress toward this goal. The first reports collected by Parliament in 1860 made it appear that tolls indeed were coming in and the turnpikes were on their way out.[8] All signs confirmed that the free market was doing its job.

CONCLUSION

Cheap local roads were on their way, thanks to the abolition of centralized government.

The immediate effect of local road management was to transfer the burden of tolls onto poor people on the periphery. In the first few decades of its success, libertarian localism dismantled the British economy, generating widespread outcry and resistance from the nation's troubled edge.

As early as 1822, signs of economic tension could be seen in the British landscape. In that year, London radicals banded together to abolish turnpikes, clamoring for roads freely used by the poor. Ostensibly, it was a victory for cheap roads and for the poor. Libertarian localism had won the day.

The cheap roads had to be paid for somehow, however. When the City of London consolidated the turnpikes, it generated an enormous sinking fund that was gradually paid off by monies collected at the remaining turnpike gates. These gates circled the outer edge of greater London. Individuals who lived just outside the area designated for tollgate abolition now paid higher tolls to accommodate the toll-free area inside the cordon.[9] The abolition of tolls stretched out in a widening belt around London in the 1830s and 1840s and shifted the burden of road payment from locals to outsiders.[10] A public debate erupted over whether the tollgate would remain at Hyde Park or be moved to Grosvenor Place. West End landholders were outraged at the suggestion of having to pay for easy access to their own homes. Kensington and the West End banded together, and the *Times* joined in, referring to "that nuisance, the toll-gate" in a series of articles.[11] Residents of Kensington, now charged many times more than they had been months ago, submitted petitions for relief. But the activists labored in vain. It was only the residents of Kensington who cared; the more numerous citizens of London were content to have outsourced responsibility. Elsewhere, attempts to extend free roads were more successful. Other petitioners managed to push the gate past Knightsbridge to the very edge of the city, securing free passage between London and Pimlico.[12] The London lobby added the criterion that one gate at some distance from the city of London should replace every six now standing, accommodating Kensington but further penalizing the neighborhoods beyond.[13] Westminster, Finsbury, Marylebone, the Tower Hamlets, and the villages of the north all still collected tolls. So piecemeal was the progress that by 1857, Londoners clamored for a mandatory, centralized assimilation of the remaining parishes into a single, open body that would take over the entirety of the area within six miles of Charing Cross, but this measure too was defeated by the localist cause.[14] Local abolition of tolls tended to leave behind such arbitrary and uneven pockets wherever a metropolitan area was too poor to pay for roads out of its coffers or unwilling for other reasons to collaborate with another body.

The costs were steeper still among poor parishes trapped in remote areas and unable to find richer parishes to take on their debts. These issues predominated in the rural parishes of Scotland and the north, where competition with foreign grain was already threatening local markets and parish coffers. The Select Committee of 1864, cognizant of such jurisdictions, recommended the forced consolidation of trusts at a local level and, again, the mandatory establishment of local highway districts in which to place these trusts.[15] New legislation in the 1860s and 1870s extended half a dozen turnpikes' charters every year, occasionally allowed bridges to become county bridges, and set up ten- or twenty-year systems for the abolition of tolls.[16] These parishes found themselves trapped in a dance of turnpike extension, never able to accumulate enough cash to pay off their tolls and unable to find government or neighboring authorities willing to help them out of their dilemma.

The coming of rail only deflected more of the burden for roads onto the residents of poor areas. William Pagan, writing as early as 1845, warned that the price of tolls would escalate in poor areas when faced with competition from the rail. In the era of road communication, long-distance travelers supplemented local road revenue. In the era of railroads, however, only locals paid for local roads. The *Edinburgh Review* summed up the problem: "The tolls formerly levied for the benefit of the neighborhood upon the passing stranger, now fall almost exclusively upon the neighborhood itself; and toll-gates, however well selected their position, cannot but operate partially and unjustly upon a purely local traffic."[17] The result was that toll roads needed stiff fares to maintain roads in anything like their former state.

Toll roads began to squeeze short-distance travelers. Turnpikes compensated for their shortfall by "placing toll-bars near railway stations."[18] In some places, Pagan found a tollhouse "for every six or eight miles of road," and the constant staffing required to maintain these tollhouses ate up the majority of the toll.[19] The trend was particularly vivid in areas like the north of Scotland where toll revenue was scarcely able to cover roads as it was.[20] In a system of local government, where roads competed with rail, the residents of poor areas found themselves paying greater and greater sums as the price of maintaining contact with the rest of the national market.

Poor parishes faced deep debts that turnpikes had accumulated under the expectation of being able to pay them off. The 1840 commission found shocking and surprising levels of unpaid debt in the trusts. Because of competition with railroads, turnpike trusts faced rising levels of unpaid interest (£821,586 in 1829; £1,123,623 in 1838) and a rapidly increasing debt overall, which had grown from £7,304,803 to £8,345,267 between 1829 and 1838. Expenses for repairs and management were barely met, and eighty-two trusts in England and Wales were unable to repay interest to their investors at all.[21] At this rate, the commissioners could hardly expect the turnpikes to repay their debts and work their way to natural extinction without government intervention. They regretfully recalled the 1836 committee's urging of "the expediency of abolishing tolls" but concluded that "under existing circumstances, they cannot with safety be dispensed with." They explained: "That it is an expensive and inconvenient method of levying money cannot be doubted, but the abolition of tolls can only be affected by a substitution of a sufficiently productive tax; such a tax we have reason to believe would be partial in its operation, and would fail to afford general relief."[22]

By the 1860s, the debts of poor districts were nearing a crisis. Reformers suggested supplementing the revenues of poor districts from the tolls of wealthier ones nearby, but even these regional toll trusts were insufficient to equalize the cost of roads.[23] Poor regions simply could not afford the cost of maintaining contact with the rest of the nation on their own. In rural areas, the debts of turnpike trusts mounted; a survey by the 1864 committee found that they owed their lenders £760,000 in interest annually. The result of these combined forces—debt, arbitrary enforcement, high toll rates falling on those least able to pay, and confusing legislation—was public outrage among the members of the poorer parishes so afflicted.

Faced with these costs, poor areas had no choice but to abandon their roads altogether. As early as the 1840s, surveyors testified in Parliament that parish roads, five years earlier the domain of expert engineers, had been left in the hands of unpaid volunteers.[24] The *Edinburgh Review* reported on parishes where the roads had deteriorated to "deeply-cut ruts" in the mud, where holes were "filled up with treacherous sand" or even "just repaired with brickbats and broken bottles."[25] The turn toward local government had caused the immediate and precipitous decline of local infrastructure.

The era of local government was marked by the persistence of turnpikes in poorer areas of the country, the decay of infrastructure around the nation, and the escalation of regional inequality. Resistance to subsidizing local revenue effectively crippled parish efforts at turnpike reform, especially in poor areas encumbered with excessive levels of turnpike debt. Localism thwarted the progress of infrastructure and allowed roads to crumble in the nation's peripheries and agricultural districts. Roused by impassioned pleas for a return to parish politics, localism flourished as an interpretation of history, a theory of administration, and a political movement, with devastating consequences for the nation's infrastructure. By the end of the nineteenth century, political localism had achieved the abandonment of Britain's roads to high tolls and infrastructural deterioration, thus reversing the trajectory of progress pioneered a century before.

Faced by mounting tolls and decaying infrastructure, the first successful centralization lobby since the 1820s took off. The road reform movement appeared in the 1860s and 1870s in rural regions where poor farmers suffered the heaviest burden of these tolls. Farmers in Salisbury cried out for a final abolition of turnpikes and collected accounts of tolls so costly that grain could not be brought to market.[26] Road reform was "much in the air" and "was debated at great length" in Aberdeenshire at the meeting of the Royal Northern Agricultural Society. James Barclay, a member of the town council and local businessman, spoke with "the breath of inspiration," moving "even" the "city folks."[27]

Road reformers called on centralized government to take over their roads once more. In the 1860s, the failing farmers looked with jealousy on the surpluses in the Consolidated Fund, the public moneys that included hereditary Crown revenues, various taxes, and other branches of revenue, now paid into the bank of England and in theory available for public works.[28] Could the fund be ransacked to save the roads? George Jenkinson came up with a plan for the Consolidated Fund to take up the remaining debts of poor localities.[29] Jenkinson, MP for Berkeley, Gloucestershire, made the rounds of provincial hotels and spoke with farmers at their annual meetings. He framed the issue in terms not only of poor localities but also of class. Jenkinson argued that only a centralized measure could overcome the logic of class in the provinces of small farmers, and that only a centralized fund would equalize "the burdens of the country" on "all classes of property."[30] The Staffordshire meeting called the reorganization of districts "a failure," explained that the expense of turnpikes was mounting, and linked them to the fact that "the roads were in a worse state than they were under the old system."[31]

Road reformers developed a new critique of local government, arguing that the nation, not the county, must be the basis for administration. Under their influence, Fabian socialists like Sidney and Beatrice Webb envisioned Britain's destiny as a path from smaller to greater bodies of political unity. The abuses of the seventeenth-century parish, they argued, had been superseded at last by the county in 1864; when the county was made obsolete by the nation, justice would prevail.

Energetic organizing at length won the day against libertarian localism. Scotland was freed from tolls in 1883, and thanks to the aggressive legislation of the committee, there were only two trusts left in 1890; the last was abolished in 1896. Money had to come from somewhere, and the last abolitions could take place only with additional aid to the counties directly from the Exchequer. Britain once again was a centralized nation, where rich regions paid for the infrastructure of poor ones.[32] In the decades that followed, centralized government returned to the rest of Britain. Regional disparities were driven home in the 1920s and 1930s as J. B. Priestly and other journalists unmasked the effects of deindustrialization in Tyneside and other areas of the north. It was as much this concern for poor regions as it was that for poor people that drove the expansion of the welfare state in Britain.

At length, the last turnpikes disappeared. Prime Minister Arthur Balfour described his vision for "great highways" uniting the nation in 1900. The Trunk Roads Act of 1936 nationalized the great arteries again for the first time since 1835. Twentieth-century Britain reimagined itself as a zone of economic connectivity made possible by infrastructure. A century after Parnell published his vision, the roads had finally become free to all.

In the early twentieth century, centralized infrastructure scored victories on an international level and came to dominate the globe. The highway system in Britain finally began to expand in the direction of something like an integrative network again with the opening of the first section of the M1 in 1959. Similar trends were explored around the world. The Italian Autostrade and the German Autobahn astonished the world in the 1920s. After the Second World War, the Netherlands launched a

CONCLUSION

similar network. Canada's state highways expanded, and it abolished its toll roads in the 1970s and 1980s. The 1956 Federal Highway Act in the United States spread roads from coast to coast; by 1989, only 5,000 of the 4 million miles of road in the United States charged tolls. In the year of the fall of the Berlin Wall, it seemed that localism had collapsed alongside communism before the final victory of a centralizing, infrastructure state.

Nineteenth-century libertarians well understood what so much state-directed building meant for political participation. Experts distanced the power of ordinary citizens and expanded their own territory. For the next hundred years, civil engineers and the experts who followed them reflected the privilege of the rich over the poor. New experts, patterning themselves on the describing, surveying civil engineers, described and mapped urban disease, crime, and poverty. From 1840 to 1910, the persuasive powers of new experts in public health, poverty, and urban planning gave them extended control over social life.

When infrastructure moved to the city, urban clearance projects were sped by the creation of new powers over property and the first implementation of eminent domain. As early as 1815, highways connecting the new post office in London were used to clear the slum of St. Martin's-le-Grand.[33] In the 1860s, massive slum clearance began in London for the planting of railway stations. Georges-Eugène Haussmann's boulevards and Kingsway-Aldwych, immense slum-clearance projects of the late nineteenth century, were sold to the public as infrastructure projects designed to better connect the city.[34] Even more influential than the civil engineers was the pattern of expert government that characterized urban reform movements in their wake. Soon after the decline of British roads, the outbreak of cholera persuaded Parliament to support another massive undertaking in the form of public sanitation. Here it was not regional interest that drove the powers of experts, but rather class interest masked as evangelical fervor. Edwin Chadwick, the sewers' architect, had been persuaded by a wave of Scottish theological pessimism of the inevitable decline of civilization through the afflictions of alcohol, dependence, and disease. In the understandings of theologians, it was the working classes who were most vulnerable to these forms of moral failure.[35] Working-class life required expert regulation to survive the coming plagues of immorality and disease. Justifying their arguments with maps and descriptions much like those of the engineers, the experts in the public health movement called for a new, centralized bureaucracy of local inspectors with powers over sewers and slaughterhouses.[36]

In the British colonies, the powers of sanitary experts had even more extended consequences. In cities like Calcutta and Johannesburg, nineteenth-century fears of contagion led by 1900 to the establishment of new laws strictly delimiting the territory where the city's ethnic inhabitants could live. By 1910, the segregated city had set the rule for the increasing disparity in economic, educational, and political availabilities for blacks and whites for the rest of the twentieth century.[37] Further extensions of urban design to control the behavior of individuals appeared in the hands of private corporations and associations. Thomas Telford's designs for the Highland Society included not only roads but also plans for ideal fishing villages that would be planted along the coast of Scotland.[38] Titus Salt's mills in Saltaire housed workers in regular, gridded dormitories and cottages where public bathhouses were forbidden but reading rooms were provided.[39] Historians have summarized these uses of housing under the heading "welfare capitalism." Through the patronage of mills, mines, and lumber companies, the design of housing became the site of social engineering projects planned to control the ordinary economic and cultural freedoms of workers across North America.[40] Legislation about housing ensured social engineering on an even broader level. Racial covenants and lending restrictions more explicitly forbade ethnic individuals from owning property in affluent neighborhoods.[41] Zoning and historic-preservation laws served the interests of the middle classes by constraining homeownership to those willing to abandon drink, those who did not trade from their own home, and those willing to keep their buildings within a certain limited aesthetic appearance.[42] These mechanisms combined to shape cities where neighborhoods were racially segregated without the force of law but rather through the power of numerous local regulations.

Expert power grew even more interventionist through the application of urban planning to social control. Early housing acts in the United Kingdom designated "problem families" and excluded them from council housing. Residents of council housing quickly became disengaged from the political alliances of the neighborhood and dedicated their time instead to family leisure.[43] The combined systems of public housing, education, and welfare were aimed at the personalities of the young. Public education and the welfare system created a rubric of a "model citizen" and the criminalizing of the "juvenile delinquent" who did not submit to his place in the social hierarchy. If he refused to believe that public school courses were serving his needs, if he refused to "queue" (stand in line) for a job, indeed, if he "loitered" among like-minded youth in a setting dedicated to popular music, he would first be refused opportunities for work and then would be targeted by the police.[44] Those who conformed survived; those who pursued their own way were cut out. In Paris, this displacement was consciously targeted as a means of breaking up zones of radical politics.[45] In the United States, the new interstate highway system almost always shot through the middle of poor, ethnic neighborhoods, breaking apart neighborhoods and school districts.[46] The clearances were justified, in the minds of their designers, by a broader plan for dispersing those communities, tagged by elites as the wellsprings of crime, poverty, and disease. As in London and Paris, infrastructure became an opportunity for social engineering. In the era of the infrastructure state, experts alone decided who could afford to participate in the market, and who had access to the circulation of the post, newspapers, and ideas.

Experts brought about an infrastructure revolution in modern cities, entailing lighting, sanitation, parks, playgrounds, subways, and utilities that followed state-led projects of road, canal, and rail. It was in these wider cases of the application of expert power to everyday life that social engineering found its full expression. In modern cities, the coming of infrastructure was rapidly followed by the birth of urban planning, the creation of public housing, and the administration of public welfare. As the tools of expert governance—surveying, abstracting, and centralization—were applied to other domains, the same issues of participation were felt across the domain of everyday life. Public health experts and urban planners came to mediate the destinies of entire neighborhoods. Each of these developments exponentially increased the opportunities for the extension of expert power into social control.

At the end of the twentieth century, however, Britain and America rehearsed again the episodes of the 1840s and 1850s. Another libertarian revolution, hostile to the state, offered its critique of experts and assembled arguments for the devolution of control to local government and private toll roads. History was repeating itself.

In Britain and the United States, the beginnings of this trend originated with the free-market revolution of the 1980s. Influenced by antistate idealism in the writings of Friedrich Hayek and Milton Friedman, the free-market regimes of Margaret Thatcher and Ronald Reagan inaugurated a long-term period of devolution from state-provided infrastructure. Leftist radicals joined their voices with the Right. In the late 1970s, for instance, MIT planning professor Robert Goodman reviewed grassroots movements against slum clearance. Along with the residents he interviewed, Goodman concluded that bureaucratic elites were bulldozing neighborhoods whose economic, social, and political unity was essential to their inhabitants' survival. He proposed a radically more open vision. "A local community should simply be able to hire or fire their own planners," he wrote, "and not have to accept them just because they were provided by the city."[47] Goodman applauded the protesters from poor communities who held "TO HELL WITH URBAN RENEWAL" signs against the bulldozers, advertising them as the sign of a reform movement to come that would reject urban planners as mediators between people and their own territory. Today, democratic activists have envisioned the Internet as the cure for political corruption, academic elitism, and environmental crisis, to name only a few domains of expertise. Like the English radicals William Cobbett and Josiah Toulmin Smith, today's radicals charge that experts, deracinated from the populations they serve, have jeopardized the nature of democracy itself.

In Britain, neoliberalism from the right prosecuted the breakup of train lines. Further measures were hinted at as well. In 2006, Britain's Parliament commissioned an overview of forward-looking options in which it highlighted the possibility of a breakup of the national highway system into privately funded turnpikes. The traveler of 2055, it forecast, would stop a dozen times between London and Edinburgh to pay separate entities their due, just as his ancestor had done in the seventeenth century.[48] At the local level, radicals on the left led the charge. London's congestion charge was greeted as a boon by environmentalists, cyclists, and urban planners, all looking forward to a city designed against the automobile. The charge reversed the work of the anti-toll-roads movement of the 1830s by permanently excluding those unable to pay from driving in the city.

Among American progressives, infrastructure issues have been overshadowed by other causes that seemed, at first glance, opposed to state-driven development. Environmentalists advocated a recentering of investment from the interstate highway system toward green infrastructure like bike lanes and light rail, typically funded by local rather than federal purses. As immigrants and poor minorities are forced by affluent gentrification to the

CONCLUSION

cities' edge, cars, roads, and highways make it possible for poor people to have a living. Commuting from home and biking to work are simply unfeasible for a majority of poor people in those situations. Local activism has ensured that bike lanes are provided to serve regional constituencies, typically favoring those affluent white neighborhoods that can afford new infrastructure over the provision of bike lanes and safe walkways to impoverished neighborhoods. Where infrastructure is concerned, green politics has traditionally elevated the ideals of the few over the hunger of the many.

In the United States, the conflicts of the Right and Left together entailed the precipitate deterioration of infrastructure. State highways in poorer regions like Michigan were merely left to decay till commuters cited "basketball-sized potholes" the length of their journeys. Amid the decaying bridges and rotting roads of Pittsburgh and Detroit, progressives planted community gardens and bicycled over gravel tracks. The collapse of neglected levees during Hurricane Katrina caused the massive evacuation of New Orleans and rendered poor people and ethnic minorities permanent refugees in trailer camps across the South. During the following years, dramatic collapses of bridges, dams, and highways demonstrated how widespread infrastructure decay had become. A nadir was reached in 2007, amid high-profile bridge collapses, when the American Society of Civil Engineers concluded that the United States was annually $1 trillion behind in maintaining its infrastructure. In cities like Cleveland and Flint, mayors elected to tear up streets and sewers rather than pay for their continued maintenance. Each of these decisions reflected complicated local and national ideas of correct expenditure.

Barack Obama's election in 2008 signaled a turn toward maintenance rather than deterioration, but reluctance to use government moneys for the cause of expansion still remained. Although the Recovery Act provided $1 trillion for road widening and repair, it created no new systems; it merely contributed money to locations where infrastructure had already been built and neglected. Running through the final decades of the twentieth century was a common thread: a reluctance to implement federal infrastructure money on behalf of growth.

In many places, long-term devolution has resulted in a broadening of the gap between rich and poor neighborhoods and regions. If poor black families are isolated in South Side Chicago, those left behind in Detroit, without jobs and grocery stores or social connections, are the most ghettoized of all. Jobs have fled that region; skyscrapers are empty; residential neighborhoods have been abandoned, subjected to arson, and left to turn into empty fields. The poor who remain are economically isolated. They depend on welfare and convenience stores for survival. There are no buses to link them to jobs, if there were any nearby, and no Internet or public libraries to tell them of opportunities outside their sphere.

The path of devolution is not the only one available in the twenty-first century. In China, by contrast, massive government expenditure has caused the overnight creation of massive cities, connected by bullet trains running at speeds up to two hundred miles an hour. Convinced that infrastructure is necessary for growth, China suffers from none of the libertarian qualms of the United States and United Kingdom. In Beijing and Shanghai, the infrastructure state is seeing a revival.

For some, a new age of information promises an escape from the tensions between local and expert control that tortured earlier generations. Digital scholar John Seely Brown describes the "optimism about the end of organization" that pervades discussions of that new zone of infrastructure, the Internet.[49] Indeed, the many-to-many connections of Internet discussion boards and e-mail lists early responded to countercultural optimism about self-governance.[50] IBM's initiatives forecast "smarter cities" where distribution problems will be settled by mass participatory platforms.[51] In the age of the Internet, old visions of participatory government are being renewed. Internet activists press for the age of the "open-source city" where public access to administration debates online will make urban government as participatory as open-source software.[52] Online surveys, public forums, and shared reports could increase public participation in debates over urban expenditure, while bus-tracking software would streamline services to particular neighborhoods.

Information does not trump infrastructure entirely, for data must flow over wires and Wi-Fi signals. A crucial mechanism of this infrastructure is the public library, that physical point of access to traditional books and computers. Never has the library been more necessary as a point of access for the poor. Few working-class individuals have reliable access; 51 percent of African Americans and 43 percent of Latinos rely on public libraries to access the Internet and learn of job opportunities.[53] As this book is being written, corporations like Microsoft and Verizon have combined forces to lobby Congress for the possibility of treating broadband access like a toll road. Their sponsors imagine a toll-road system of Internet content, monitoring users as they get on the Internet and charging them as they get off. Like the toll roads of the eighteenth century, these systems would be prone to local monopolies and inflated rates. They would tend to be cheaper in rich areas where multiple services compete to provide McMansions with swiftly downloaded videos. In enclaves of the rich, the new elite would buy and sell online; in the ghettos of poorer regions, the hardworking poor would have to wait offline for entrepreneurs to require their services. Meanwhile, advocates of state broadband, on the other hand, envision an era when poor high-school students download the classics from Google, poor mothers advertise their clerical services, and racial minorities lobby and vote online.[54]

The warnings of history are clear. The decline of connectivity today mirrors the outcome of the nineteenth-century libertarian revolution that plunged Scotland and Ireland into a century of poverty. When states back away from infrastructure, the victims are economically isolated regions and ethnicities. Without the aid of connection by infrastructure, marginalized populations become even more disconnected.

Infrastructure inevitably pits interests against one another: peripheries must be connected or collapse. In the era of new technologies, infrastructure again raises the question whether states will choose to make participatory markets or not. Affluent interests must shoulder the burden of construction and profit only indirectly from the connectivity of the poor. Regimes may choose radical segregation over integration, but infrastructure has become too necessary to modern experience to be ignored.

NOTES

ACKNOWLEDGMENTS

INDEX

NOTES

Introduction

1. Susan Sherratt and Andrew Sherratt, "The Growth of the Mediterranean Economy in the Early First Millennium BC," *World Archaeology* 24:3 (February 1993): 361–378; David Christian, "Silk Roads or Steppe Roads?" *Journal of World History* 11:1 (Spring 2000): 1–26; Jerry H. Bentley, "Cross-Cultural Interaction and Periodization in World History," *American Historical Review* 101:3 (June 1996): 749–770; William Hardy McNeill, *Plagues and Peoples* (New York: Anchor Books, 1989); Philip D. Curtin, *Cross-Cultural Trade in World History* (Cambridge: Cambridge University Press, 1984); Christopher I. Beckwith, *Empires of the Silk Road* (Princeton, N.J.: Princeton University Press, 2009); Frances Wood, *The Silk Road* (Berkeley: University of California Press, 2004); R. Foltz, *Religions of the Silk Road* (New York: Palgrave Macmillan, 2000).

2. Ray Laurence, "Afterword," in Colin Adams and Ray Laurence, eds., *Travel and Geography in the Roman Empire* (London: Routledge, 2001), 167. After the time of Augustus, the official highways of the empire allowed the emperor to travel, hear provincial petitions, and make decisions outside Rome. J. G. Landels, *Engineering in the Ancient World* (Berkeley: University of California Press, 2000); Xavier de Planhol and Paul Claval, *An Historical Geography of France* (Cambridge: Cambridge University Press, 1994); M. I. Finley, "Empire in the Greco-Roman World," *Greece and Rome* 25:1 (April 1978): 1–15; Stephen Mitchell, "Imperial Building in the Eastern Roman Provinces," *Harvard Studies in Classical Philology* 91 (1987): 333–365; Charles D. Trombold, *Ancient Road Networks and Settlement Hierarchies in the New World* (Cambridge: Cambridge University Press, 1991); David French, "Pre- and Early-Roman Roads of Asia Minor: The Persian Royal Road," *Iran* 36 (1998): 15–43; Ray Laurence, "Milestones, Communications, and Political Stability," in Linda Ellis, ed., *Travel, Communication, and Geography in Late Antiquity* (Aldershot, Hants., England: Ashgate, 2004), chap. 4; Rodney S. Young, "Gordion on the Royal Road," *Proceedings of the American Philosophical Society* 107:4 (August 1963): 348–364.

3. R. S. Lopez, "The Evolution of Land Transport in the Middle Ages," *Past and Present* 9 (April 1956): 17–29; Howard L. Adelson, "Early Medieval Trade Routes," *American Historical Review* 65:2 (January 1960): 271–287; Robert Arthur Leeson, *Travelling Brothers* (London: G. Allen and Unwin, 1979); Henri Pirenne, *Medieval Cities* (Princeton, N.J.: Princeton University Press, 1969); Colin Adams, "Getting Around in Ancient Egypt" in Adams and Laurence, *Travel and Geography in the Roman Empire*, 137–167; Planhol and Claval, *Historical Geography of France*; Barry W. Cunliffe, *Facing the Ocean* (Oxford: Oxford University Press, 2002); James Tracy, *The Rise of Merchant Empires* (Cambridge: Cambridge University Press, 1990).
4. Tim Ingold, *The Perception of the Environment* (New York: Routledge, 2000); Kenneth Olwig, *Landscape, Nature, and the Body Politic* (Madison: University of Wisconsin Press, 2002); J. A. Yelling, *Slums and Slum Clearance in Victorian London* (London: Allen and Unwin, 1986); David H. Pinkney, *Napoleon III and the Rebuilding of Paris* (Princeton, N.J: Princeton University Press, 1958); Christopher Hamlin, *Public Health and Social Justice in the Age of Chadwick: Britain, 1800–1854* (Cambridge: Cambridge University Press, 1998); Catherine J. Kudlick, *Cholera in Post-revolutionary Paris* (Berkeley: University of California Press, 1996).
5. Martha D. Pollak, *Turin, 1564–1680* (Chicago: University of Chicago Press, 1991); Chandra Mukerji, *Impossible Engineering* (Princeton, N.J.: Princeton University Press, 2009); Antoine Picon, *French Architects and Engineers in the Age of Enlightenment* (Cambridge: Cambridge University Press, 2010).
6. Olwig, *Landscape, Nature, and the Body Politic*; Jan de Vries, *Barges and Capitalism* (Wageningen: Landbouwhogeschool, 1978); H. C. Darby, *The Draining of the Fens* (Cambridge: University Press, 1956); Eric Pawson, *Transport and Economy* (New York: Academic Press, 1977); Steven C. A. Pincus, *1688* (New Haven, Conn.: Yale University Press, 2009).
7. Daniel Defoe, *A Tour thro' the Whole Island of Great Britain*, 4th ed., 4 vols. (London: Printed for S. Birt, 1748 [1724–1727]), 1:334.
8. Maxine Berg, *The Age of Manufactures, 1700–1820* (London: Routledge, 1994).
9. Stephen Daniels and Denis Cosgrove, *The Iconography of Landscape* (Cambridge: Cambridge University Press, 2002); Emma Rothschild, *Economic Sentiments* (Cambridge, Mass.: Harvard University Press, 2001).
10. Rosalind Williams, "Cultural Origins and Environmental Implications of Large Technological Systems," *Science in Context* 6:2 (1993): 377–403; Michael Heffernan, "On Geography and Progress: Turgot's *Plan d'un ouvrage sur la géographie politique* (1751) and the Origins of Modern Progressive Thought," *Political Geography* 13:4 (July 1994): 328–343; Jessica Riskin, "The 'Spirit of System' and the Fortunes of Physiocracy," in Margaret Schabas and Neil De Marchi, eds., *Oeconomies in the Age of Newton* (Durham, N.C.: Duke University Press, 2003), 42–73.

11. Clay McShane, *The Horse in the City* (Baltimore: Johns Hopkins University Press, 2007).
12. Matthew Simons, *A Direction for the English Traviller* (London: Printed and are to be sold by John Garrett, at the South entrance of ye Royall Exchange in Corn-hill, [1677?]); John Ogilby, *Britannia* (London: Printed by the author, 1675). See also discussion of maps in Chapter 4, notes 178–179.
13. *Bowles's New Traveller's Guide* (London: Carington Bowles, 1777); *Bowles's New Pocket Guide* (London: Carington Bowles, 1780); Daniel Paterson, *A New and Accurate Description of all the Direct and Principal Cross Roads* (London: Daniel Paterson, 1781); Frederick Ebenezer Baines, *On the Track of the Mail-Coach: Being a Volume of Reminiscences Personal and Otherwise* (London: Richard Bentley and Son, 1895), 118; George Fordham, "The Work of John Cary and His Successors," *Geographical Journal* 63:5 (May 1924): 438–440; Herbert George Fordham, *John Cary, Engraver, Map, Chart and Print-Seller and Globe-Maker, 1754–1835: A Bibliography with an Introduction and Biographical Notes* (Cambridge: University Press, 1925).
14. Carington Bowles, *Post Chaise Companion* (London: Carington Bowles, 1782), G. F. Cruchley, *Cruchley's New Travelling Map and Itinerary for England* (London: G. F. Cruchley, 1831), and John Betts, *Betts's New Railway and Commercial Map of England and Wales* (London: John Betts, 1846). See Chapter 4, note 182.
15. Defoe, *Tour thro' the Whole Island of Great Britain*, 1:257, 264.
16. Ibid., 1:257, 366, 374; 2:114.
17. Dan Bogart, "Did Turnpike Trusts Increase Transportation Investment in Eighteenth-Century England?" *Journal of Economic History* 65 (2005): 439–468.
18. Defoe, *Tour thro' the Whole Island of Great Britain*, 1:175, 368.
19. The various acts and committees that dealt with the progress of the military roads in Scotland, the Highland Roads and Bridges, and the Holyhead Road will be documented extensively in this book. Secondary sources include Sidney Webb and Beatrice Webb, *English Local Government: The Story of the King's Highway* (London: Longmans, Green, 1913); Ian L. Donnachie, *Roads and Canals, 1700–1900* (Edinburgh: Holmes-McDougall, 1977); R. B. Haldane, *New Ways through the Glens: Highland Road, Bridge and Canal Makers of the Early Nineteenth Century* (Colonsay: House of Lochar, 1995); A. J. Quartermaine, B. Trinder, and R. Turner, *Thomas Telford's Holyhead Road* (York: Council for British Archaeology, 2003); and Frank Goddard, *The Great North Road: A Guide for the Curious Traveller* (London: Frances Lincoln, 2004).
20. "Turnpike Roads and Bridges," *Hansard's Parliamentary Debates* (May 15, 1860), 1301. The figure for parliamentary investment dates from 1826, the high-water point of parliamentary expenditure. "Account of Sums Advanced to

Commissioners of Holyhead Roads, Holyhead and Howth Harbours, and Menai and Conway Bridges, 1825–26," *House of Commons, Parliamentary Papers,* 1826 (355). By 1848, investment was reduced to £16,509. Commissioners of Woods, Forests, and Land Revenues, Works and Buildings, *Twenty-sixth Report,* 1849 (611).

21. Select Committee on Holyhead and Liverpool Roads, *Report,* 1830 (432).
22. Robert Southey, "On the Accounts of England by Foreign Travelers," in *Essays, Moral and Political* (London: John Murray, 1832), 265, quoting Charles Moritz, *Travels in England in 1782* (London: Printed for G. G. and J. Robinson, 1795), 12.
23. John Loudon McAdam, *Remarks on the Present System of Road Making* (London: Printed for Longman, Hurst, Rees, [1816] 1824), 7, 38. Macadam spelled his name Macadam, M'Adam, and McAdam; I have adopted the rule of referring to him as Macadam (more prevalent in the parliamentary papers) except where his name is printed otherwise.
24. 5 & 6 Will. 4, c. 50 (General Highway Act, 1835).
25. Defoe, *Tour thro' the Whole Island of Great Britain,* 1:6–7, 18–19, 71.
26. W. T. Jackman, *The Development of Transportation in Modern England* (Cambridge: University Press, 1916), 157–210, 255–451; Philip S. Bagwell, *The Transport Revolution* (London: Routledge, 1974), 1–21.
27. Thomas Cox, *Magna Britannia et Hibernia, Antiqua & Nova; or, A New Survey of Great Britain,* 6 vols. (London: Nutt and Morphew, 1720–1731). The same was true of Herman Moll, *The Compleat Geographer; or, The Chorography and Topography of All the Known Parts of the Earth* (London: J. Napton, 1723). A similar absence of roads characterized most seventeenth-century maps and charts of Britain.
28. William Camden, *Britannia; or, A Chorographical Description of Great Britain and Ireland, together with the Adjacent Islands,* 2 vols. (London: James and John Knapton, 1722).
29. Eugen Weber, *Peasants into Frenchmen* (Stanford, Calif.: Stanford University Press, 1976); Frederic L. Paxson, "The Highway Movement, 1916–1935," *American Historical Review* 51:2 (January 1946): 236–253.
30. H. Robinson, *The British Post Office: A History* (Princeton, N.J.: Princeton University Press, 1948); Eugène Vaillé, *Histoire des postes françaises jusqu'à la Révolution* (Paris, 1957); David M. Henkin, *The Postal Age* (Chicago: University of Chicago Press, 2006).
31. Sidney Webb and Beatrice Webb, *English Local Government: The Story of the King's Highway* (London: Longmans, Green, 1913); G. R. Taylor, *The Transportation Revolution, 1815–1860* (New York: Rinehart, 1951); H. J. Dyos and D. H. Aldcroft, *British Transport: An Economic Survey from the Seventeenth Century to the Twentieth* (Leicester: Leicester University Press, 1969); William Albert, *The Turnpike Road System in England, 1663–1840* (Cambridge: University

Press, 1972); Eric Pawson, *Transport and Economy: The Turnpike Roads of Eighteenth Century Britain* (London: Academic Press, 1977); R. Szostak, *The Role of Transportation in the Industrial Revolution* (Montreal: McGill–Queen's University Press, 1991).

32. T. S. Ashton, *The Industrial Revolution, 1760–1830* (New York: Oxford University Press, 1948); David S. Landes, *The Unbound Prometheus* (London: Cambridge University Press, 1969); Joel Mokyr, *The Economics of the Industrial Revolution* (Totowa, N.J.: Rowman and Allanheld, 1985); N. F. R. Crafts, *British Economic Growth during the Industrial Revolution* (New York: Oxford University Press, 1985).

33. Commissioners of Highland Roads and Bridges, *Statement of Origin and Extent of Roads in Scotland, and Papers Relating to Military Roads, 1813–1814* (63), 17, 34.

34. Select Committee on Survey and Report of Coasts and Central Highlands of Scotland, *Report*, 1802–1803 (80).

35. Christopher Bailey, "Making and Meaning in the English Countryside," in David E. Nye, ed., *Technologies of Landscape: From Reaping to Recycling* (Amherst: University of Massachusetts Press, 1999), 140; Russ Haywood, *Railways, Urban Development and Town Planning in Britain: 1948–2008* (Farnham: Ashgate, 2009), 42–43.

36. John M. Barry, *Rising Tide* (New York: Simon and Schuster, 1997); David Blackbourn, *The Conquest of Nature: Water, Landscape, and the Making of Modern Germany* (New York: Norton, 2006); Theodore M. Porter, *Trust in Numbers* (Princeton, N.J.: Princeton University Press, 1995).

37. N. G. Onuf and P. S. Onuf, *Nations, Markets, and War* (Charlottesville: University of Virginia Press, 2006); Edward Fox, *History in Geographic Perspective: The Other France* (New York: Norton, 1972); Robin L. Einhorn, *Property Rules* (Chicago: University of Chicago Press, 2001).

38. Margaret C. Jacob, "The Cosmopolitan as a Lived Category," *Daedalus* 137:3 (July 1, 2008): 18–25; P. Kleingeld and E. Brown, "Cosmopolitanism," in *Stanford Encyclopedia of Philosophy* (2006).

39. For a general history of the transition from way finding to route planning, then under way with the coming of new kinds of geographic tools, see Catherine Delano-Smith and Roger Kain, *English Maps: A History* (Toronto: University of Toronto Press, 1999), 168. See also Catherine Delano-Smith, "Milieus of Mobility: Itineraries, Route Maps and Road Maps," in James Akerman, *Cartographies of Travel and Navigation* (Chicago: University of Chicago Press, 2006), chap. 2.

40. Defoe, *Tour thro' the Whole Island of Great Britain*, 2:361.

41. Samuel Simpson, *The Agreeable Historian; or, The Compleat English Traveller*, 2 vols. (London: R. Walker, 1746), 2:208; *Domestic Life in England, from the*

Earliest Period to the Present Time (London: Thomas Tegg and Son, 1835), 252–253.
42. Defoe, *Tour thro' the Whole Island of Great Britain*, 3:211, 1:18, 74.
43. The minimum widths were changed by 11 Geo 3, c. 53, sects. 1–3. See George Tait, *A Summary of the Powers and Duties of a Justice of the Peace in Scotland* (Edinburgh: Anderson, 1821), 153; John Tuke, *General View of the Agriculture of the North Riding of Yorkshire* (London: Printed for G. Nicol, 1800), 303.
44. Defoe describes four major variations on the Great North Road, where the main artery of England's commerce crossed the soft marly soil of the Midlands: one to York via Royston and Stilton; one through St. Albans, Dunstable, Northampton, and Nottingham or Darby; another through Coventry and West Cheshire; and another through Worcester. Defoe, *Tour thro' the Whole Island of Great Britain*, 1:331.
45. William Kitchiner even advised travelers on how to tell a good inn from a bad inn, warning against "certain houses" where "certain state beds" were "kept for certain Visitors" that were "very likely to be damp." William Kitchiner and John Jervis, *The Traveller's Oracle; or, Maxims for Locomotion* (London: Henry Colburn, 1827), 116. Jervis provided a list of fifteen ways of distinguishing a good driver from a bad one. Such advice books provided a defense against the charges of collusion between innkeepers and stagecoach drivers or boat keepers, which still allowed innkeepers in remote areas to bleed their guests. See James Playfair, *A Geographical and Statistical Description of Scotland* (Edinburgh: Printed for A. Constable, 1819), 362; Joshua Toulmin, *The History of Taunton* (London: Printed by T. Norris, sold by J. Johnson, 1791), 604.
46. See the discussion of consulting with the innkeeper in Priscilla Wakefield, *A Family Tour through the British Empire* (London: Darton, Harvey and Darton, 1816), 141.
47. Anne Macvicar Grant, *Letters from the Mountains: Being the real correspondence of a lady, between the years 1773 and 1803* (Printed for Longman, Hurst, Rees, and Orme, 1806), 12–13.
48. William Albert, "Popular Opposition to Turnpike Trusts in Early Eighteenth-Century England," *Journal of Transport History* 5:1 (February 1979): 1–17; Phillip D. Jones, "The Bristol Bridge Riot and Its Antecedents: Eighteenth-Century Perception of the Crowd," *Journal of British Studies* 19:2 (1980): 74–92; Miles Ogborn, *Spaces of Modernity* (New York: Guilford Press, 1998); Humphrey R. Southall, "The Tramping Artisan Revisits: Labour Mobility and Economic Distress in Early Victorian England," *Economic History Review* 44:2 (1991): 272–296; Christopher Otter, "Cleansing and Clarifying: Technology and Perception in Nineteenth-Century London," *Journal of British Studies* 43:1 (2003): 40–64; Humphrey R. Southall, "Agitate! Agitate! Organize! Political

Travellers and the Construction of a National Politics, 1839–1880," *Transactions of the Institute of British Geographers* 21:1 (1996): 177–193.

1. Military Craft and Parliamentary Expertise

1. John Rickman and Samuel Smiles give different versions of the story. John Rickman, "Introduction," in Thomas Telford, *The Life of Thomas Telford, Civil Engineer, Containing a Descriptive Narrative of His Professional Labours* (London: Payne and Foss, 1838), vi–viii; Samuel Smiles, *The Life of Thomas Telford, Civil Engineer* (London: J. Murray, 1867), 309.
2. Rickman, "Introduction," 1:vi–viii.
3. Tess Canfield, "Edmund Turrell," in A. W. Skempton, ed., *A Biographical Dictionary of Civil Engineers* (London: Thomas Telford, 2002); Telford, *Life*; G. Watson, *The Civils* (London: Thomas Telford, 1988).
4. The original diagram of the ideal road section had been prepared in 1824 by another engraver, Aaron Arrowsmith, who also contributed engravings to Telford's autobiography and, like Turrell, was elected to the Institute of Civil Engineers on Telford's nomination. Tess Canfield, "Aaron Arrowsmith," in Skempton, *Biographical Dictionary of Civil Engineers*, 1:21–22; "Arrowsmith," in *A New General Biographical Dictionary*, 12 vols. (London: B. Fellowes, Ludgate Street, 1853), 2:210.
5. Edmund Turrell, "An Improved Mode of Forming and Sharpening the Points of Etching-Needles and Dry-Points," *Gill's Technical Repository* 2 (1822): 254–259; Edmund Turrell, "On the Necessity of Employing Pure Nitrous Acid, in Etching Copper-Plates," *Technical Repository* 3 (1823): 55.
6. John Timbs, *Stories of Inventors and Discoverers in Science and the Useful Arts: A Book for Old and Young* (New York: Harper and Brothers, 1860); John Timbs, *School-Days of Eminent Men* (London: Lockwood, 1862); Samuel Smiles, *Lives of the Engineers* (London: J. Murray, 1861); Smiles, *Life of Thomas Telford*; Grant Allen, *Biographies of Working Men* (London: Society for Promoting Christian Knowledge, 1885); James Parton, *Captains of Industry* (Boston: Riverside Press, 1896); William Henry Maxwell, *The Construction of Roads and Streets* (London: St. Bride's Press, 1899); Alexander Gibb, *The Story of Telford: The Rise of Civil Engineering* (London: A. Maclehose and Co., 1935); Roy Devereux, *John Loudon McAdam: Chapters in the History of Highways* (London: Oxford University Press, 1936); L. T. C. Rolt, *Thomas Telford* (London: Longmans, Green, 1958); Keith Ellis, *Thomas Telford, Father of Civil Engineering* (London: Priory Press, 1974); Derrick Beckett, *Telford's Britain* (Newton Abbot, Devon; North Pomfret, Vt.: David and Charles, 1987). Contemporary scholars are beginning to challenge this myth: Antoine Picon, "Engineers and Engineering History: Problems and Perspectives," *History and Technology* 20:4

(2004): 421–436; Martin Barnes, "Thomas Telford, Project Manager," *Civil Engineering* 160 (2007): 61–64; James Anderson, "Urban Development as a Component of Government Policy in the Aftermath of the Napoleonic War," *Construction History* 15 (1999): 23–37; Dana Arnold, "Rationality, Safety and Power: The Street Planning of Later Georgian London," *Georgian Group Journal* 5 (1995): 37–50.

7. T. S. Ashton, *The Industrial Revolution, 1760–1830* (New York: Oxford University Press, 1948); G. R. Taylor, *The Transportation Revolution, 1815–1860* (New York: Rinehart, 1951); W. T. Jackman, *The Development of Transportation in Modern England* (London: Frank Cass, 1966); H. J. Dyos and D. H. Aldcroft, *British Transport: An Economic Survey from the Seventeenth Century to the Twentieth* (Leicester: Leicester University Press, 1969); David S. Landes, *The Unbound Prometheus: Technological Change and Industrial Development in Western Europe from 1750 to the Present* (London: Cambridge University Press, 1969); William Albert, *The Turnpike Road System in England, 1663–1840* (Cambridge: University Press, 1972); Eric Pawson, *Transport and Economy: The Turnpike Roads of Eighteenth Century Britain* (London: Academic Press, 1977); Benedict Anderson, *Imagined Communities: Reflections on the Origin and Spread of Nationalism* (London: Verso Editions/NLB, 1983); Ernest Gellner, *Nations and Nationalism* (Ithaca, N.Y.: Cornell University Press, 1983); Joel Mokyr, *The Economics of the Industrial Revolution* (Totowa, N.J.: Rowman and Allanheld, 1985); N. F. R. Crafts, *British Economic Growth during the Industrial Revolution* (New York: Oxford University Press, 1985); Maxine Berg, *The Age of Manufactures: Industry, Innovation, and Work in Britain, 1700–1820* (New York: Oxford University Press, 1986); Rick Szostak, *The Role of Transportation in the Industrial Revolution: A Comparison of England and France* (Montreal: McGill–Queen's University Press, 1991); Linda Colley, *Britons: Forging the Nation, 1707–1837* (New Haven, Conn.: Yale University Press, 1992); Dan Bogart, "Turnpike Trusts and the Transportation Revolution in 18th Century England," *Explorations in Economic History* 42:4 (2005): 479–508.

8. John Brewer, *The Sinews of Power: Money, War, and the English State, 1688–1783* (Boston: Unwin, 1989); Lawrence Stone, *An Imperial State at War: Britain from 1689 to 1815* (New York: Routledge, 1994); Philip Harling and Peter Mandler, "From 'Fiscal-Military' State to Laissez-Faire State, 1760–1850," *Journal of British Studies* 32:1 (1993): 44–70; Gianfranco Poggi, *The Development of the Modern State* (Stanford, Calif.: Stanford University Press, 1978); T. Skocpol, *States and Social Revolutions: A Comparative Analysis of France, Russia and China* (Cambridge: Cambridge University Press, 1979); Gianfranco Poggi, *The State: Its Nature, Development, and Prospects* (Stanford, Calif.: Stanford University Press, 1990); Charles Tilly, *Social Movements, 1768–2004* (Boulder, Colo.: Paradigm Publishers, 2004).

9. George Wade, *Report* (1725), quoted in *The Roads and Railroads, Vehicles, and Modes of Travelling, of Ancient and Modern Countries* (London: John W. Parker, 1839), 171.
10. James Dorret, *A General Map of Scotland and Islands Thereto Belonging from New Surveys* (London, 1750); Burt, *Letters from a Gentleman in the North of Scotland to His Friend in London*, 2 vols. (London: Printed for S. Birt, in Ave-Maria Lane, 1754); *The Life of John Metcalf, Commonly Called Blind Jack of Knaresborough* (York: E. and R. Peck, 1795); *Roads and Railroads*, 174; George Anderson and Peter Anderson, *Guide to the Highlands and Islands of Scotland*, 3rd ed. (Edinburgh: Adam and Charles Black, 1851), 45; J. B. Salmond, *Wade in Scotland* (Edinburgh: Moray Press, 1934); William Taylor, *The Military Roads in Scotland* (Newton Abbot: David and Charles, 1976); "General Wade," in Skempton, *Biographical Dictionary of Civil Engineers*, 751–753.
11. Burt, *Letters from a Gentleman*, 2:304–305.
12. Ibid., 2:327.
13. Contemporary treatises on the art of perspective offered little or no commentary on mountains and other perspectives of complex distance. Bernard Lamy, *A Treatise of Perspective; or, The Art of Representing All Manner of Objects' as They Appear to the Eye in All Situations* (London, 1702); John Shuttleworth, *A Treatise of Opticks* (London: Printed for Dan. Midwinter, 1709); Humphry Ditton, *A Treatise of Perspective Demonstrative and Practical* (London: B. Tooke and D. Midwinter, 1712); Brook Taylor, *Linear Perspective; or, A New Method of Representing Justly All Manner of Objects as They Appear to the Eye in All Situations* (London: R. Knaplock, 1715).
14. Jean Dubreuil, *The Practice of Perspective; or, An Easy Method of Representing Natural Objects According to the Rules of Art* (London: Printed for Tho. Bowles and John Bowles, 1726), 11.
15. Burt, *Letters from a Gentleman*, 2:10, 310.
16. Burt, *Letters from a Gentleman*, 1:328.
17. R. A. Skelton, "The Military Survey of Scotland, 1747–1755," *Scottish Geographical Magazine* 83:1 (1967): 1–15.
18. Adam Martindale, *The Country Survey-Book; or, Land Meter's Vade-Mecum* (London: Printed for R. Clavel and G. Sawbridge, 1702), i.
19. Edward Laurence, *The Young Surveyor's Guide* (London: Printed for James Knapton, 1716), title page.
20. George Adams, *Geometrical and Graphical Essays* (London: Printed by R. Hindmarsh, 1791).
21. Henry Wilson, *Surveying Improv'd; or, The Whole Art, Both in Theory and Practice, Fully Demonstrated* (London: Printed for J. Batley, 1743), 521.
22. Sarah Bendall, "Estate Maps of an English County: Cambridgeshire, 1600–1836," in David Buisseret, ed., *Rural Images: Estate Maps in the Old and New*

Worlds (Chicago: University of Chicago Press, 1996), 63–90; Peter Eden, "Land Surveyors in Norfolk, 1550–1850. Part 1: The Estate Surveyors," *Norfolk Archaeology* 35 (1973): 474–482; W. R. Ward, "The Administration of the Window and Assessed Taxes, 1696–1798," *English Historical Review* 67:265 (1952): 522–542.

23. Paul Sandby, chief draughtsman of the military survey of Scotland, and his brother Thomas Sandby, one of the military surveyors, went on to lives as civilian surveyors, directing the construction of Virginia Water. Thomas Sandby became architect of the king's works in 1777.

24. Bruce Robertson, *Paul Sandby and the Early Development of English Watercolor* (New Haven, Conn.: Yale University Press, 1987).

25. Nicolas Bion, *The Construction and Principal Uses of Mathematical Instruments* (London: Printed by H. W. for John Senex, 1723). Such works were a radical advance over early instruction manuals such as those of Edward Laurence and Henry Wilson, mentioned earlier, and S. W., *The Practical Surveyor; or, The Art of Land-Measuring, Made Easy* (London: Printed for J. Hooke and J. Sisson, Mathematical Instrument-Maker, 1725).

26. The British military had access to Mandey's *Marrow of Measuring* (1710), the most sophisticated statement of using the chain to ascertain distances accurately. See also Edmund Stone Cunn, *Treatise of the Construction and Use of the Sector* (London: Printed for John Wilcox, 1729), which described charting territory with a chain or circumferenter and measuring angles with a theodolite or compass, and William Gardiner, *Practical Surveying Improved; or, Land-Measuring, According to the Present Most Correct Methods* (London: J. Sisson, 1737), which laid out the principle of the "new-invented Spirit-Level." Many of these works demonstrate their authors' affiliation with the military: see Paul Hoste, *A Compendious Course of Practical Mathematicks, Particularly Adapted to the Use of the Gentlemen of the Army and Navy*, 3 vols. (London: Printed for A. Bettesworth and C. King, 1730), by the military instructor of naval officers, which was dedicated to the Duke of Argyle, the master-general of the ordnance.

27. Yolande Jones, "Aspects of Relief Portrayal on 19th Century British Military Maps," *Cartographic Journal* 9:1 (1974): 19–33.

28. William Roy, *An Account of the Measurement of a Base on Hounslow-Heath, by Major-General William Roy, Read at the Royal Society, from April 21 to June 16, 1785* (London: Printed by J. Nichols, 1785); William Mudge, *An Account of the Operations Carried On for Accomplishing a Trigonometrical Survey of England and Wales; From the Commencement, in the Year 1784*, 3 vols. (London: Printed by W. Bulmer and Co. for W. Faden, 1799–1811); Edward Williams, *An Account of the Trigonometrical Survey Carried On in the Years*

1791, 1792, 1793, and 1794, by Order of His Grace the Duke of Richmond (London: The Royal Society, 1795).
29. George Adams, *Geometrical and Graphical Essays*; Luke Hebert, *The Engineer's and Mechanic's Encyclopaedia* (London: T. Kelly, 1837).
30. John Hammond, *The Practical Surveyor* (London: Heath, 1731); Thomas Breaks, *A Complete System of Land-Surveying, Both in Theory and Practice, Containing the Best, the Most Accurate, and Commodious Methods of Surveying and Planning of Ground by All the Instruments Now in Use* (Newcastle upon Tyne: Printed by T. Saint for W. Charnley, 1771); Charles Hutton, *The Compendious Measurer; Being a Brief, yet Comprehensive, Treatise on Mensuration and Practical Geometry* (London: G. G. and J. Robinson, 1786); Charles Hutton, *Elements of Conic Sections; With Select Exercises in Various Branches of Mathematics and Philosophy, for the Use of the Royal Military Academy at Woolwich* (London: Printed for J. Davis, 1787); William Garrard, *Copious Trigonometrical Tables, Shewing the Results in All Cases of Plane Trigonometry, by Inspection* (London: J. Moore, 1789).
31. George Adams, *Geometrical and Graphical Essays*.
32. For instance, Hutton; see note 30 above. A driving force behind these innovations, of course, was maritime surveying: Murdoch Mackenzie, *A Treatise on Maritim Surveying* (London: Printed for Edward and Charles Dilly, 1774). Benjamin Donn's research on geometry was likewise applied to maritime surveying. See Benjamin Donn, *A New Introduction to the Mathematicks* (London: Printed for W. Johnston, 1758).
33. Edmund Gunter's four-part chain, originally pioneered in the seventeenth century, appeared in common use in the 1770s and purported to free surveyors from what Arthur Burns called the "errors of estimation that unavoidably attend the scale and protractor." Arthur Burns, *Geodaesia Improved, or a New and Correct Method of Surveying Made Exceedingly Easy* (Chester: J. Poole, 1771), title page. Popularized by William Hume and Arthur Burns, the new method of measuring with the chain emphasized calculating the diagonal distance of any quadrangle and using trigonometry to double-check the figures put down for the quadrangle's length and breadth. See also William Nicholson, "Gunter, (Edmund)," *The British Encyclopedia, or, Dictionary of Arts and Sciences*, vol. 3.
34. Gardiner, *Practical Surveying Improved*.
35. Eden, "Land Surveyors in Norfolk, 1550–1750. Part 1"; and Peter Eden, "Land Surveyors in Norfolk, 1550–1850. Part 2," *Norfolk Archaeology* 36 (1975): 119–148.
36. Bendall, "Estate Maps," 5, 10. Increasingly, surveyors were not only useful but also mandatory. After parliamentary approval, any act of enclosure, turnpike building, canal building, town surveying, or paving required the drafting of

surveyors. Meanwhile, the laws instituting the assessed taxes and window taxes of 1696 and 1747 directed a small army of surveyors to every house in the nation; a revision of the code in 1785 expanded the number of surveyors employed to 32 in Middlesex and 132 in the surrounding counties. See also Ward, "Administration of the Window and Assessed Taxes."

37. I. H. Adams, "Economic Process and the Scottish Land Surveyor," *Imago Mundi* 27 (1975): 13–18. Bendall, "Estate Maps," 44–46, includes details for Wales and Ireland as well.
38. R. A. Buchanan, *The Engineers: A History of the Engineering Profession in Britain, 1750–1914* (London: Jessica Kingsley, 1989), 38–39, 45.
39. Gardiner, *Practical Surveying Improved*.
40. Thomas Dix, *A Treatise on Land-Surveying, in Six Parts* (London: Printed by H. Baldwin and Son, 1799).
41. Early eighteenth-century records might verbally describe the bounds of an estate or delineate property boundaries with a few lines. These maps were of no use to road builders operating by sighting. Surveys instead were first deployed in an administrative context, where they became requisite for parties making any sale or alteration to their property in order that parish authorities might be able to better establish the relationship between changing routes. In Norfolk, maps typically accompanied the surveyor's affidavit for enclosure legislation by the 1790s and became legally required in 1815. As late as 1798, the surveyor general reported of royal properties that "leases were granted, frequently, without having Surveys taken of the Property, and that the Surveys, when taken, were never confirmed by the Oath of the Surveyor, nor any other means employed for ascertaining the Value of the Estates which could be attended with Expence, no Fund being provided for defraying that charge." Eden, "Land Surveyors in Norfolk, 1550–1850. Part 2," 127. See also *The First Report of the Surveyor General of His Majesty's Land Revenue; in Obedience to the Directions of an Act of Parliament* (London: Land Revenue Office, 1798).
42. John Sykes, *A Chronology, by Year, Month and Day, from* A.D. 120 *to Oct. 9, 1832* (London: T. Fordyce, 1832), 60.
43. John Gwynn, *London and Westminster Improved* (London: Sold by Dodsley, 1766).
44. John Holt, *General View of the Agriculture of the County of Lancaster: With Observations on the Means of Its Improvement* (London: G. Nicols, 1795), 196, original emphasis.
45. Arthur Young, *A Six Weeks Tour through the Southern Counties of England and Wales* (London: W. Nicoll, 1768), 219.
46. Henry Law and Samuel Hughes, *Rudiments of the Art of Constructing and Repairing Common Roads* (London: J. Weale, 1850), 13.

47. *Life of John Metcalf*, 133.
48. Quoted in Smiles, *Life of Thomas Telford*, 90.
49. *Life of John Metcalf*, 128.
50. Smiles, *Life of Thomas Telford*, 86.
51. Burt, *Letters from a Gentleman*, 1:339; 2:305–308.
52. Smiles enumerates the following routes: In Yorkshire: between Harrogate and Boroughbridge; between Harrogate and Harewood Bridge; Chapeltown to Leeds, Broughton to Addingham, Mill Bridge and Halifax; Wakefield and Dewsbury; Wakefield and Doncaster; Wakefield, Huddersfield, and Saddleworth (the Manchester Road); Standish and Thurston Clough; Huddersfield and Highmore; Huddersfield and Halifax; and Knaresborough and Weatherby. In Lancashire: the Bury and Blackburn; Blackburn and Accrington; Haslingden and Accrington. Between Yorkshire and Lancashire: Skipton, Colne, and Burnley; Dock-Land Head and Ashton-under-Lyne; Alston to Stockport; and Stockport to Mottram Langdale. In Cheshire and Derby; the Macclesfield and Chaple-en-le-Frith, the Whaley and Buxton; and the Congleton and the Red Bull. Smiles, *Life of Thomas Telford*, 89.
53. Henry Sacheverell Homer, *An Enquiry into the Means of Preserving and Improving the Publick Roads of This Kingdom* (Oxford: Printed for S. Parker, 1767).
54. *Life of John Metcalf*, 129, 135.
55. Ibid., 146.
56. Ibid., 138–140.
57. Æ, *An Essay for the Construction of Roads on Mechanical and Physical Principles* (London: Printed for T. Davies, 1774), 46.
58. Joseph Hodgkinsen, *The Farmer's Guide; or, An Improved Method of Management of Arable Land* (London: Printed for the author, 1794), 32, 34. The same formula was repeated by John Middleton, *View of the Agriculture of Middlesex* (London: G. Nicol, 1798), 400. Both explicitly recommended the language of foundations, drainage, and exactly broken stones "the size of a hen's egg" that would be repeated by Macadam a decade later.
59. John Sinclair, *The Statistical Account of Scotland, Drawn Up from the Communications of the Ministers of the Different Parishes*, 21 vols. (Edinburgh: W. Creech, 1791–1799), 14:84; *Communications to the Board of Agriculture: On Subjects Relative to the Husbandry, and Internal Improvement of the Country*, 7 vols. (London: Printed by W. Bulmer for G. and W. Nicol, 1797–1813), 1:144.
60. James Brindley, *Reports by James Brindley Engineer, Thomas Yeoman Engineer, and F. R. S. and John Golborne Engineer, Relative to a Navigable Communication betwixt the Firths of Forth and Clyde, Edinburgh, 13th, 23d, 30th September, 1768* (Edinburgh: Printed by Balfour, Auld, and Smellie, 1768), 40; John

Ferrar, *The History of Limerick, Ecclesiastical, Civil and Military, from the Earliest Records, to the Year 1787* (Limerick: A. Watson and Co., 1787), 87; James Brindley, *The History of Inland Navigations, Particularly That of the Duke of Bridgwater* (London: T. Lowndes, 1766), 79; Ralph Dodd, *Report on the First Part of the Line of Inland Navigation from the East to the West Sea by Way of Newcastle and Carlisle, as Originally Projected* (Newcastle: Printed for the Subscribers, 1795), 46.

61. Homer, *Enquiry into the Means of Preserving and Improving the Publick Roads*, 30.
62. *Schemes Submitted to the Consideration of the Publick, More Especially to Members of Parliament, and the Inhabitants of the Metropolis* (London: Printed for W. Browne, 1770), 51–52.
63. John Smith, *General View of the Agriculture of the County of Argyll, with Observations on the Means of Its Improvement* (Edinburgh: Printed by Mundell, 1798), 276–277.
64. Lord Middleton caused a cut on the Derby road near Nottinghamshire around 1740. Robert Thoroton, *The Antiquities of Nottinghamshire, Extracted out of Records, Original Evidences, Leiger-Books, Other Manuscripts, and Authentic Authorities* (Nottingham: Printed by G. Burbage, 1790), 4; John Tuke, *General View of the Agriculture of the North Riding of Yorkshire, with Observations on the Means of Its Improvement* (London: Printed by W. Bulmer, 1794), 84. The leveling of roads formed a major part of the advice promoted by the Board of Agriculture in the 1790s in *Communications to the Board of Agriculture*, 1:143 and passim.
65. In 1807, when the commissioners of the Highland Roads and Bridges contracted with Archibald Campbell, a Scottish surveyor, to make their roads, they set out specifications that did not differ much from the kind of foundation technology that Metcalf had learned from General Wade's road builders almost a century before. Campbell was instructed to lay a foundation of stones not "above the size of a hen's egg" to the depth of four inches and then cover them with another ten inches of gravel. In bogs and moors, he was supposed to clear the moss, if there was a stone surface less than two feet beneath; if the underbed was too deep, he was to float the road on "swarded turf, heather, or brushwood." In sighting the road, he was instructed "to cut down all Heights, fill up all Hollows, and blow all the Rocks." Coms. of Highland Roads and Bridges, *Third Report*, 1807 (100), 18, 52–53.
66. Arthur Young, *A Six Weeks Tour, through the Southern Counties of England and Wales* (London: Printed for W. Strahan, 1768), 372–380. Young's tours also include *The Farmer's Tour through the East of England*, 4 vols. (London: Printed for W. Strahan, 1771).
67. Quoted in Hebert, *Engineer's and Mechanic's Encyclopædia*, 601.

68. John Houghton, *Husbandry and Trade Improv'd: Being a Collection of Many Valuable Materials Relating to Corn, Cattle, Coals, Hops, Wool, &c.* (London: Printed for Woodman and Lyon, 1727).
69. Arthur Young noted these variations and compared them negatively with the military gravel paths built by improvers. Young, *Six Weeks Tour, through the Southern Counties*, 1:88, 145.
70. Most of Young's journeys took him over parish roads managed by private turnpike trusts. He found evidence both of turnpikes that truly improved road conditions and of those that merely charged travelers for a profit. Arthur Young, *Six Weeks Tour, through the Southern Counties*, 1:120–121, 153, 305, 323; 2:28.
71. Sighting monuments were stones of prodigious size, lifted "by the Force of Engines and Strength." Burt, *Letters from a Gentleman*, 1:290. For a description of the process, see also Thomas Pennant, *A Tour in Scotland* (Chester: John Monk, 1771), 1:214.
72. "General Wade's Speech" and "Sir William Wyndam's Speech," in *A Collection of the Parliamentary Debates in England, from the Year MDCLXVIII to the Present Time*, 21 vols. (London: 1739–1742), 10:87–88; *The Historical Register, Containing an Impartial Relation of All Transactions* 12 (London, 1727), 2, 78–79.
73. "General Wade," 751–753; *Roads and Railroads*, 171–174.
74. *Life of John Metcalf*, 124, 138.
75. William George Maton, *Observations Relative Chiefly to the Natural History, Picturesque Scenery, and Antiquities, of the Western Counties of England* (Salisbury: Printed by J. Easton, 1797), 33; see also William Marshall, *The Rural Economy of the West of England, Including Devonshire; and Parts of Somersetshire, Dorsetshire, and Cornwall* (London: Nicol, 1796), 194.
76. Hodgkinsen, *Farmer's Guide*, 35.
77. The Turnpike Acts are as follows: 5 Geo. 1, c. 12; 26 Geo. 2, c. 30; 30 Geo. 2, c. 28; 6 Geo. 3, c. 43; and 7 Geo. 3, c. 40 and c. 42. See also A. B., *Some Observations on the Use of Broad Wheels: With a Proposal for a More General Amendment of the Roads* (London, 1765); Alexander Cumming, *Observations on the Effects Which Carriage Wheels, with Rims of Different Shapes, Have on the Roads* (London, 1797); Robert Anstice, *Remarks on the Comparative Advantage of Wheel Carriages of Different Structure and Draught* (London: Printed for the author by S. Symes, 1790); and Edward Kenney, *A Memoir on Wheel Carriages* (Cork: Printed by A. Edwards, 1792).
78. 26 Geo. 2, c. 30, sect. 15, explained, "Every owner of every wagon, wain, or cart, traveling for hire, driving or causing to be driven, drawn, or conveyed such wagon, &c., upon any turnpike road, shall have his Christian and Surname and Place of Abode, written or painted in large legible letters

upon the Tilt, or other conspicuous place of such wagon." 28 Geo. 2, c. 17, sect. 14 provided the same measures for all narrow-wheeled stage wagons or carts. 18 Geo. 2, c. 33, sect. 4, explained, "No person whatsoever shall drive any cart, car, or dray, of any kind whatsoever, within the limits of the weekly bills of mortality, unless the master or owner of such cart, shall place upon some conspicuous part of such cart the name of the owner, and the number of such cart so belonging to him, in order that the driver of such cart, &c. may be the more easily convicted for any disorder or misbehaviour committed by him." Sect. 5: "Every owner of such cart residing within the limits aforesaid, shall enter his name and place of abode with the commissioners for licensing Hackney Coaches, for which entry no more than one shilling shall be paid."

79. 28 Geo. 3, c. 57 (1788); 30 Geo. 3, c. 34 (1790).
80. William Felton, *A Treatise on Carriages: Comprehending Coaches, Chariots, Phaetons, Curricles, Whiskeys, &c.* (London: Printed for and sold by the author, 1794); Charles Wilks, *Observations on the Height of Carriage Wheels* (Cork, 1814); James Walker, *Observations on the Wheels of Carriages* (Edinburgh: William Blackwood, 1824); William Deacon, *Remarks on Conical and Cylindrical Wheels, Public Roads, Wheel Carriages &c.* (London: Printed for the author, 1808); Timothy Sheldrake, *A Description of the Theory and Properties of Inclined Plane Wheels: By Which Power and Velocity Will Be Increased, and Friction Diminished* (London: Printed for Egerton, 1811); Richard Lovell Edgeworth, *A Letter to the Dublin Society, Relative to Experiments on Wheel Carriages* (Dublin: Printed by Graisberry and Campbell, 1816); Richard Lovell Edgeworth, *An Essay on the Construction of Roads and Carriages* (London: Printed for J. Johnson, 1813); T. Fuller, *An Essay on Wheel Carriages* (London: Printed for Longman, Rees, Orme, Brown, and Green, 1828); Alice Newlin, "An Exhibition of Carriage Designs," *Metropolitan Museum of Art Bulletin* 35:10 (1940): 185–191; Desmond King-Hele, "Erasmus Darwin's Improved Design for Steering Carriages—and Cars," *Notes and Records* 56:1 (January 2002): 41–62.
81. "The Examination of Mr. William Clarke," appendix 1 in Select Committee on Acts Regarding Use of Broad Wheels, and Regulations for Preservation of Turnpike Roads and Highways of United Kingdom, *First Report*, 1806 (212), 6.
82. 27 Geo. 3, c. 13. The act pertained only to the duties of customs, excise, stamps, and expenses. It was broadened to include total revenues in 1803 and the pay of all public servants in 1810 (50 Geo. 3, c. 117), which brought all other offices into the game of annual accounting, to the great protest of the disorganized and secretive Post Office. Norman Chester, *The English Administrative System, 1780–1870* (Oxford: Clarendon Press, 1981), 99–100.

83. Select Committee on Acts Regarding Use of Broad Wheels, and Regulations for Preservation of Turnpike Roads and Highways of United Kingdom, *Second Report*, 1806 (321). Another 135-page report was published in 1808.
84. John Loudon McAdam, *Remarks on the Present System of Road Making* (London: Longman, Hurst, Rees, [1816] 1824), 215. Macadam spelled his name Macadam, M'Adam, and McAdam; I have adopted the rule of referring to him as Macadam (more prevalent in the parliamentary papers) except where his name is printed otherwise.
85. "Estimates, Miscellaneous Services," *House of Commons Sessional Papers*, 1839, *Reports of Commissioners* (142-I-V), 8.
86. See Henry Parnell's indictment of Macadam, Macadam's indictment of John Wingrove's system, and James Paterson's vicious attack on Macadam: Henry Parnell, *A Treatise on Roads* (London: Longman, Rees, Orme, Brown, Green and Longman, 1833), 74–75; McAdam, *Remarks on the Present System of Road Making*, 215; James Paterson, *A Series of Letters and Communications Addressed to the Select Committee of the House of Commons on the Highways of the Kingdom* (London: Montrose, 1822), 5–8. Other proposals include Edgeworth, *Essay on the Construction of Roads and Carriages*; J. J. A. MacCarthy, *Prospectus of a New Pavement for the Carriage-Ways of the Metropolis* (1815); John Loudon McAdam, *Memorial on the Subject of Turnpike Roads* (1818); T. C., *Road-Making on Mr. M'Adam's System* (Edinburgh, 1819); John Loudon McAdam, *A Practical Essay on the Scientific Repair and Preservation of Public Roads* (London: B. McMillan, 1819); James Paterson, *A Practical Treatise on the Making and Upholding of Public Roads* (Montrose: for T. Donaldson, 1819); A. H. Chambers, *Observations on the Formation, State and Condition of Turnpike Roads and Other Highways* (London: Printed for the author, 1820); *Hints to Country Road-Surveyors* (London, 1820); Bryan Donkin, *A Paper, Read before the Institution of Civil Engineers, on the Construction of Carriage-Way Pavements* (London: J. Taylor, 1824); William Deykes, *Considerations on the Defective State of the Pavement of the Metropolis* (London: J. Taylor, 1824); J. Finlayson, "Improved Method of Paving Streets," *Glasgow Mechanics' Magazine and Annals of Philosophy* 73 (Saturday, May 14, 1825): 226–228; "Plan of Making Roads," *Glasgow Mechanics' Magazine and Annals of Philosophy* 66 (Saturday, April 2, 1825): 113–118; Charles Penfold, *A Proposed Amendment in the Highway Laws of England Addressed to the Members of Both Houses of Parliament* (Croydon: J. S. Wright, 1834); Andrew Henderson, *Practical Hints on the Principles of Surveying and Valuing Estates: Also, on the Making, Repairing, and Management of Roads* (Edinburgh: Ballantyne, 1829); "Biddle's Machine for Repairing Roads," *Newton's Journal* 13 (1827): 27–28; Thomas Fall, *The Surveyor's Guide; or, Every Man His Own Road Maker: Comprising the*

Whole Art of Making and Repairing Roads, Prices for Work, Forming of Estimates, and Office of Surveyor (Retford: Printed for the author by F. Hodson, 1827).

87. Select Committee on Acts Regarding Use of Broad Wheels, *Second Report*, 1808 (225), 7n.
88. The political origins and success of the lobby for highways are the subject of Chapter 2 of this book.
89. Select Committee on Acts Regarding Use of Broad Wheels, *First Report*, 1806 (321), 45.
90. John Young, *The Improvement of the Marsh, and of the Country Near about It* (London, 1701); John Smeaton, *Reports of the Late John Smeaton Made on Various Occasions, in the Course of His Employment as a Civil Engineer* (London: Longman, Hurst, Rees, Orme, and Brown, 1812); review, "Reports of the Late John Smeaton, F. R. S., Made on Various Occasions in the Course of His Employment as a Civil Engineer," *Gentleman's Magazine*, March 1813, 245; Review, "Reports of the Late John Smeaton, F. R. S., Made on Various Occasions in the Course of His Employment as a Civil Engineer," *Eclectic Review*, January–June 1813, 53; Richard Lovell Edgeworth, "On the Construction of Theatres," *Journal of Natural Philosophy, Chemistry, and the Arts* 23 (1809): 129; John Rennie, *Address of Sir John Rennie, President, to the Annual General Meeting of the Institution for Civil Engineers* (London: Published at the Institution, 1846).
91. "Introduction," in Edgeworth, *Essay on the Construction of Roads and Carriages*; "Engineer, civil," in Thomas Martin, *The Circle of the Mechanical Arts; Containing Practical Treatises on the Various Manual Arts, Trades, and Manufactures* (London: R. Rees, 1813), 293–316.
92. Telford, *Life of Thomas Telford*, 1:279.
93. R. L. Edgeworth and M. Edgeworth, *Memoirs of Richard Lovell Edgeworth* (London: R. Hunter, 1820); R. L. Edgeworth and M. Edgeworth, *Practical Education*, 2 vols. (London: J. Johnson, 1798); R. E. Schofield, *The Lunar Society of Birmingham* (Oxford: Clarendon Press, 1963); D. Clarke, *The Ingenious Mr. Edgeworth* (London: Osborne, 1965).
94. Edgeworth, *Essay on the Construction of Roads and Carriages*.
95. Roy Devereux, *John Loudon McAdam*; R. H. Spiro, "John Loudon McAdam, Colossus of Roads" (PhD diss., University of Edinburgh, 1950); W. J. Reader, *Macadam: The McAdam Family and the Turnpike Roads, 1798–1861* (London: Heinemann, 1980).
96. The "General Rules" first appeared in the 1811 Report of the Committee on Parliamentary Roads, appendix C, 31–32. It was afterwards republished and distributed in McAdam's numerous publications: McAdam, *Remarks on the Present System of Road Making* (1816); J. L. McAdam, *Observations on the*

Management of Trusts for the Care of Turnpike Roads (London: Printed for Longman, Hurst, Rees, Orme, Brown, and Green, 1825).

97. Telford, *Life of Thomas Telford*; Smiles, *Lives of the Engineers*; Gibb, *Story of Telford*; Rolt, *Thomas Telford*; A. Alastair Penfold, *Thomas Telford, "Colossus of Roads"* (London: Telford Development Corporation, 1981); Henry Parnell, *A Treatise on Roads* (London: Longman, Rees, Orme, Brown, Green and Longman, 1833); C. Hadfield, *Thomas Telford's Temptation* (Cleobury Mortimer, Shropshire: M. and M. Baldwin, 1993); A. R. B. Haldane, *New Ways through the Glens* (London: Thomas Nelson, 1962).

98. Mr. Farey, "Observations Regarding the Formation of Roads," appendix 9 in Select Committee on Acts Regarding Use of Broad Wheels, *First Report*, 1806 (212), 51.

99. Select Committee on Acts Regarding Use of Broad Wheels, *First Report*, 1808 (225), 5, 107–117.

100. The military form was already a popular structure among administrators; Edgeworth admired the Scottish roads and suggested them for Dublin when he was mayor there. But no one before had described military methods as precisely as did Macadam. The road would rest on a convex "bottoming" with "twelve inches of broken stones" in layers of four inches, each battered down by a period of passing traffic. Above the bottoming would rest a two-inch bed of "fine gravel . . . cleansed from all earth." Quantitative measurements and specific descriptors added an air of replicable precision to Macadam's suggestion. Thomas Telford, "Report," appendix 2 in Select Committee on Acts Regarding Turnpike Roads and Highways in England and Wales, *Report*, 1810–1811 (240), 23; Edgeworth, *Essay on the Construction of Roads and Carriages*, 29–48; Parnell, *Treatise on Roads*, 132.

101. Select Committee on Acts Regarding Turnpike Roads and Highways in England and Wales, *Report*, 1819 (509), 46–48. In the 1819 version of the Holyhead report, Telford acknowledged the benefits of using local materials. Describing the Wolverhampton Trust, he explained, "This Trust procures materials with difficulty, and at a great expense." He advised that local practices and local relationships could supply those materials at less cost than a government bureaucracy: "It would be prudent to hold out liberal terms for farmers to lay down stones by the sides of the road, at a rate per ton weight, or per cubic yard, which might be the means of bringing them from a considerable distance, on each side of the road"; indeed, "these pebble stones are much superior to any other material, that every means should be resorted to procure them; and, in order that they might be used for the top, working metal only, I would advise to pave the bottom of the road with hard free-stone, of which there is plenty to be had conveniently in the trust." Committee for Improvement of Holyhead Roads, on State of Road between London and Shrewsbury,

Report, 1820 (126), 21. Parnell too argued that geography required diverse paving. Parnell explained that "in a district of country where any coarse sort of stone can be got," it would be "cheaper" to pave the whole with flagstones than to import gravel. Parnell, *Treatise on Roads*, 155. Rejecting Macadam's "General Rules," other civil engineers preferred to make local recommendations on the basis of setting and geography.

102. John MacNeill, Telford's assistant, ran experiments with a cement foundation in the Highgate Archway section of the Holyhead Road and reported two years later that it was still "perfectly hard," differing in every way from the wreck of clay, gravel, and sand that had made the road a disaster beforehand. Parnell, *Treatise on Roads*, 131–132, 168–169. MacNeill was also interviewed by the sixth committee on the Holyhead Road. Telford and Parnell endorsed a variety of possible solutions for local trusts: wooden railways, for example, might be "preferable to canals" for set destinations in industry wherever water was scarce. Only those roads that received so much traffic "as to wear down three inches of hard broken stones in a year" should be paved. Parnell, *Treatise on Roads*, 107, 145.

103. The paving stones recommended for city streets would be rectangles of granite or, where that was unavailable, limestone or freestone. By 1830, Telford's smooth pavements had replaced Macadam's gravel as the preferred method of paving road surfaces of local roads and city streets where local governments still had the choice of their own method of pavement. Telford's alternative for intercity highways, first used in the Holyhead Road, differed from Macadam's in both surface and foundation. The surface was composed of stones of four or five inches, bound together with very fine gravel in between. The paving stones had their "irregularities . . . broken off by the hammer, and all the interstices . . . filled with stone chips firmly wedged or packed by hand with a light hammer." The foundation was composed of wider stones set flat against one another at the bottom; Parnell, *Treatise on Roads*, 131. See also Thomas Telford, "Report on Pavements," appendix no. 2 in Parnell, *Treatise on Roads*, and A. Walker, "Plan of a Species of Single Railway," appendix no. 16 in Select Committee on Acts Regarding Use of Broad Wheels, *First Report*, 1808 (225), 125. This model was composed of broken stone from its foundation to its upper surface. Ibid., 148, 157, 160.

104. The first-class roads would have paving stones ten inches thick and fifteen by eight inches wide. "Streets of the second class" and "third class" would have stones eight and six inches deep, respectively. Parnell, *Treatise on Roads*, 107, 131. These categories derived from the text of a French engineer of the Ponts et Chaussées, Joseph Sganzin, who defined in his *Elementary Course of Civil Engineering* roads of four classes, ranging from the highways that connect "the capitals of states" or "large cities" down through major turnpikes running between cities, connective county roads, and town roads. In Parnell's usage,

the distinction was really between parliamentary, first-class roads and connective parish roads. Joseph Mathieu Sganzin, *An Elementary Course of Civil Engineering* (Boston, Mass.: 1837; originally Paris, 1809), 76; cf. the distinction taken up by Charles Le Mercher de Longpré, Baron de Haussez, *Great Britain in 1833* (London: R. Bentley, 1833), 127.

105. James Loudon McAdam, *Observations on the Highways of the Kingdom* (London, 1816); John Loudon McAdam, *Memorial on the Subject of Turnpike Roads* (1818); T. C., *Road-Making on Mr. M'Adam's System*; John Loudon McAdam, *A Practical Essay on the Scientific Repair and Preservation of Public Roads* (London: B. McMillan, 1819).

106. 5 Geo. 4, c. 100; 6 Geo. 4, c. 38; Parnell, *Treatise on Roads*, 139.

107. Select Committee on the Roads from Holyhead to London, *First Report*, 1817 (313), 7; Parnell, *Treatise on Roads*, 146.

108. 55 Geo. 3 required a deep foundation with a broken stone surface for all new roads.

109. Coms. for Improvement of Holyhead Roads, on State of Road between London and Shrewsbury, *Report*, 1820 (126), 11, 40.

110. Thomas Telford, "Report," appendix 2 in Select Committee on State of Roads from Shrewsbury to Holyhead, and from Chester to Holyhead, *Report*, 1810–1811 (197), 23.

111. Thomas Telford, "Survey and Report of Coasts and Central Highlands of Scotland," *House of Commons Sessional Papers*, 1802–1803, *Reports of Commissioners* (45), 3–4.

112. Ibid. See also Thomas Telford, "Report and Estimate on Two Proposed Lines of Road in N. England," *House of Commons Sessional Papers*, 1820, *Reports of Commissioners* (279), 2; Telford, "Reports upon the State of Road between London and Shrewsbury," Committee for Improvement of Holyhead Roads, 1820 (126); Telford, "Reports on the State of the Road from London to Liverpool," Committee for the Liverpool and London Road, *Report*, 1826–7 (362); Telford, "Reports, Estimates, and Plans for Improving Road from London to Liverpool," Committee for the London and Liverpool Roads, 1829 (123); Act for a New Road at St. Alban's, 7 Geo. 4, c. 74; Committee for Improvement of Road from London to Holyhead, and from London to Liverpool, *Fifth Report*, 1828 (476); Select Committee on Report, Plan and Estimate of Road from Carlisle to Glasgow, *Report*, 1814–1815 (463).

113. Select Committee on State of Roads from Shrewsbury to Holyhead, and from Chester to Holyhead, *Report*, 1810–1811 (197).

114. Select Committee on Holyhead Road and Harbour, *Second Report*, 1810 (352), 13.

115. Parnell, *Treatise on Roads*, 45–46, 48.

116. Ibid., 79.

117. Telford, "Report and Estimate," 6.

118. Telford, "Reports on the State of the Road from London to Liverpool," Committee for the Liverpool and London Road, *Report*, 1826–1827 (362), 4.
119. John Benjamin MacNeill, *Tables for Calculating the Cubic Quantity of Earth Work in the Cuttings and Embankments of Canals, Railways, and Turnpike Roads* (London: Roake and Varty, 1833).
120. Parnell, *Treatise on Roads*, 45.
121. Select Committee on Acts Regarding Use of Broad Wheels, *First Report*, 1808 (225), 33.
122. Parnell, *Treatise on Roads*, 172; Select Committee of House of Lords on Fees for Turnpike Road Bills, *Second Report*, 1833 (703), 137; "Roads," in Hebert, *Engineer's and Mechanic's Encyclopædia*, 605; "Road Indicator," *Magazine of Popular Science* 1 (1836): 140.
123. John MacNeill, "Description of the Road Indicator," appendix no. 1 in Parnell, *Treatise on Roads*, 326–327.
124. "Highways," in *The Farmer's Encyclopædia and Dictionary of Rural Affairs* (London: Longman, Brown, Green, and Longmans, 1842), 631–632.
125. William Jardine, *Memoirs of Hugh Edwin Strickland, M.A.* (London: John van Voorst, 1858), 89.
126. Parnell, *Treatise on Roads*, 243. Parnell was still being quoted as an authority in his own right by popular publications nearly ten years later. Society for the Diffusion of Useful Knowledge, *Penny Cyclopaedia of the Society for the Diffusion of Useful Knowledge* (London: C. Knight, 1841), 32.
127. Select Committee on Acts Regarding Turnpike Roads and Highways in England and Wales, *Report*, 1819 (509), 23–24; Telford, *Life of Thomas Telford*, 1:xx.
128. Parnell, *Treatise on Roads*, 63–66, 69–72.
129. Commissioners for the Improvement of the Mail Coach Road from London, by Coventry, to Holyhead, *General Rules for Repairing Roads* (London: J. Taylor, 1827), 30.
130. MacNeill, "Description of the Road Indicator," 326–327.
131. Thomas Telford, "Survey and Report of Coasts and Central Highlands of Scotland," Committee for Improvement of Holyhead Roads, on State of Road between London and Shrewsbury, *Report*, 1820 (126), 5.
132. Select Committee on State of Road from London to Holyhead, by Coventry and Shrewsbury, *Sixth Report*, 1819 (549), 132–133.
133. Parnell, *Treatise on Roads*, 133.
134. Thomas Telford, "Reports to Commissioners for Improvement of Holyhead Roads," House of Commons Reports 1820 (126), 4–5, 8.
135. Parnell, *Treatise on Roads*, 153.
136. Ibid., 86–87.
137. 8 & 9 Will. 3, c. 16.
138. Parnell, *Treatise on Roads*, 179.

139. Ibid., 140–144.
140. House of Commons, "Estimate of Sum Required in 1821 for Commissioners Acting under the 55 Geo. 3 c. 152 to Complete Sundry Improvements," 1821 (633).
141. Lord Viscount Morpeth quizzing Henry Parnell, March 17, 1830, included in the Select Committee on State of the Northern Roads, *Report*, 1830 (172),14.
142. Daniel Defoe recorded numerous hulks of incomplete military projects for ports and fortresses deemed irrelevant or too expensive at the close of hostilities. Daniel Defoe, *A Tour thro' the Whole Island of Great Britain*, 4th ed., 4 vols. (London: Printed for S. Birt, 1748 [1724–1727]), 1:6–7, 18–19, 71.
143. Select Committee on Petitions and Documents for New Post Office at Cheapside, *Report*, 1813–1814 (338), appendix; "Report from the Select Committee on Intended Improvements in the Post-Office," Post Office Archive 91/3, 1815. Investigation of tenants had begun in 1814. Under examination was Joseph Hillman, an ironmonger of Foster Lane. The examining judge, William Bankes, had heard evidence that the site under discussion, St. Martin's-le-Grand, was populated by some dozen "houses of ill repute." As Post Office officials assembled a case that tearing down the entire neighborhood would benefit the public good, the court had heard the location described as a pernicious den of thieves and a menace to law-abiding society. At length the Select Committee turned to interview local tenants and householders. Hillman countered that St. Martin's parish was "valuable to the inhabitants." He spoke of silversmiths living near Goldsmith Hall and of old families and local traditions. He added that the inhabitants were for the most part old and would be greatly inconvenienced by the forced removal. "Do you think the inhabitants of St. Martin's-le-Grand are older inhabitants, more stationary, than in general the inhabitants in other parts of the city?" asked Bankes. Hillman replied, "There are not people of longer standing in any neighbourhood, than in the neighbourhood of St. Martin's-le-Grand." Were there any brothels there? asked Bankes, who had heard testimony that the entire neighborhood was nothing but shopfronts on the outside and dens of vice within. Hillman replied that there were two. Cf. the later but similar case of Regent Street: Committee on the Petition of the Tradesmen and Inhabitants of Norris Street, *Minutes of Evidence*, 1817.

2. Colonizing at Home

1. Philip Harling and Peter Mandler, "From Fiscal-Military State to Laissez-Faire State, 1760–1850," *Journal of British Studies* 32:1 (1993): 44–70; Philip Harling, "Rethinking Old Corruption," *Past and Present* 147 (1995): 127–158.

2. Boyd Hilton, *The Age of Atonement: The Influence of Evangelicalism on Social and Economic Thought, 1785–1865* (Oxford: Oxford University Press, 1991); Christopher Hamlin, "Edwin Chadwick and the Engineers, 1842–1854: Systems and Antisystems in the Pipe-and-Brick Sewers War," *Technology and Culture* 33:4 (1992): 680–709; Christopher Hamlin, *Public Health and Social Justice in the Age of Chadwick* (New York: Cambridge University Press, 1998); Christopher Otter, "Cleansing and Clarifying: Technology and Perception in Nineteenth-Century London," *Journal of British Studies* 43:1 (2003): 40–64; Patrick Joyce, *The Rule of Freedom: Liberalism and the Modern City* (New York: Verso, 2003); Gianfranco Poggi, *The Development of the Modern State* (Stanford, Calif.: Stanford University Press, 1978); Theda Skocpol, *States and Social Revolutions: A Comparative Analysis of France, Russia and China* (Cambridge: Cambridge University Press, 1979); Gianfranco Poggi, *The State: Its Nature, Development, and Prospects* (Stanford, Calif.: Stanford University Press, 1990); Christopher Tilly, *Social Movements, 1768–2004* (Boulder, Colo.: Paradigm Publishers 2004).
3. The other social experiments he envisioned included a project for a national bank, a national pension office, and an asylum for idiots. Daniel Defoe, *Essays upon Several Projects* (London: Thomas Ballard, 1702), 1; Walter Wilson, *Memoirs of the Life and Times of Daniel Defoe*, 3 vols. (London: Hurst and Chance, 1830), 1:256
4. Samson Eure, *Trials Per Pais; or, The Law of England* (London: Printed by Eliz. Nutt, and R. Gosling, 1718), 433. The Highway Act of 1734 made parish surveyors responsible for fencing off such holes in the road and filling them in or sloping them down within three days of discovering them. 7 Geo. 2, c. 42, sects. 20–22. The Highway Act of 1773 allowed for two justices of the peace to indict the digger for 20 shillings to 5 pounds. 13 Geo. 3, c. 78, sect. 33
5. Defoe, *Essays upon Several Projects*, 75.
6. William Coxe, *Travels into Poland, Russia, Sweden, and Denmark*, 2 vols. (London: T. Cadell, [1784] 1785), 2:488.
7. "Marquis de Condorcet's Life of Turgot," *Monthly Review* (1787): 631.
8. Arthur Young, *Travels during the Years 1787, 1788 and 1789*, 2 vols. (Dublin: Printed for Messrs. R. Cross, P. Wogan, L. White, P. Byrne, A. Grueber [1792] 1793), 2:511.
9. In fact, farmers could dodge statute labor by paying the surveyors a fine that was less than the cost of their daily absence from work, so something resembling such a system was in place anyway. Indeed, replacement of statute duty by fines was the path followed when the Highway Act of 1835 abolished it for good. John Shapleigh, *Highways: A Treatise Shewing the Hardships, and Inconveniences of Presenting, or Indicting Parishes, Towns, etc., for Not Repairing the Highways* (London: Printed for S. Birt, 1749); John Hawkins,

Observations on the State of the Highways (London: Printed for J. Dodsley, 1763); Richard Whitworth, *Some Remarks upon a Plan of a Bill Proposed to Parliament, for Amending the Highways* (London: J. Dodsley, 1764).
10. "Forced purchase," the British term, and "eminent domain," the American term, are used interchangeably throughout this book. No real history of the uses of eminent domain exists, although this book deals substantially with the moment when the powers of the state were first applied in the modern West to reconfiguring the structures of urban space. The history of eminent domain is bound up with the centralization of state control over urban territory. For background on the history of the laws on eminent domain, see "The Right of Eminent Domain," *American Law Register* 4:11 (September 1856): 641–661; Philip Nichols, *The Power of Eminent Domain* (Boston: Boston Book Co., 1909); G. M. W., "Eminent Domain: Validity of State Statute," *Michigan Law Review* 34:3 (January 1936): 424–426; Arthur Lenhoff, "The Development of the Concept of Eminent Domain," *Columbia Law Review* 42:4 (April 1942): 596–638; H. N. Scheiber, "Public Rights and the Rule of Law in American Legal History," *California Law Review* 72:2 (1984): 217–251; H. N. Scheiber, "The Jurisprudence—and Mythology—of Eminent Domain in American Legal History," in E. F. Paul and H. Dickman, eds., *Liberty, Property, and Government: Constitutional Interpretation before the New Deal* (Albany: State University of New York Press, 1989), 217, 223–225; Susan Reynolds, *Before Eminent Domain: Toward a History of Expropriation of Land for the Common Good* (Chapel Hill: University of North Carolina Press, 2010).
11. Defoe, *Essays upon Several Projects*, 76–77.
12. John Scott, *Digests of the General Highway and Turnpike Laws* (London, 1778).
13. Peter Barfoot, *A Candid Review of Facts, in the Litigation between Peter Barfoot, Esq. and Richard Bargus, and Others, with the Bishop of Winchester* (London: Green & Co., 1788), 7.
14. *The Case and Allegations of the Trustees of the Totnes Road-Act* (London, 1761), 2.
15. *Lounger* 78 (Saturday, July 29, 1786): 88–89.
16. By the 1830s, some nine hundred turnpike trusts managed twenty thousand miles of road in England and Wales, about 17 percent of the entire road network. William Albert, "The Metropolis Roads Commission: An Attempt at Turnpike Trust Reform," *Transport History* 4 (1971): 225–244; William Albert, *The Turnpike Road System in England, 1663–1840* (Cambridge: University Press, 1972); John Howard Chandler, *The Amesbury Turnpike Trust* (Salisbury, South Wilts.: South Wiltshire Industrial Archaeology Society, 1979); M. C. Lowe, *Turnpikes and Tollgates: A Study of the Totnes and Bridgetown Pomeroy Turnpike Trust*,

1759–1881 (Totnes: Totnes Community Archive, 1987); A. D. M. Phillips and B. J. Turton, "The Turnpike Network of Staffordshire, 1700–1840: An Introduction and a Handlist of Turnpike Acts," *Collections for a History of Staffordshire*, 4th ser., 13 (1988): 61–118; James Barfoot, "An Investigation of an Eighteenth-Century Montgomeryshire Turnpike and Its Origins," *Montgomeryshire Collections* 77 (1989): 73–80; Chris Budgen, "The Bramley and Rudgwick Turnpike Trust," *Surrey Archaeological Collections* 81 (1991–1992): 97–102; W. B. Taylor, "A History of the Tadcaster-York Turnpike," *York Historian* 12 (1995): 40–61; M. C. Lowe, "The Exeter Turnpike Trust, 1753 to 1884," *Devonshire Association Report and Transactions* 127 (1995): 163–188; W. M. Hunt, "The Promotion of Tattershall Bridge and the Sleaford to Tattershall Turnpike," *Lincolnshire History and Archaeology* 31 (1996): 41–45; Simon Morris, "The Marylebone and Finchley Turnpike, 1820–50," *Camden History Review* 21 (1998): 24–32; Dan Bogart, "Turnpike Trusts, Infrastructure Investment, and the Road Transportation Revolution in Eighteenth-Century England" (PhD diss., UCLA, 2005).

17. The best source on underfunded turnpike conditions and nonturnpiked roads remains Young, *Travels during the Years 1787, 1788 and 1789*.
18. See p. 229, n. 77 above.
19. J. C. D. Clark, *English Society, 1688–1832* (Cambridge: Cambridge University Press, 1985); J. H. Plumb, *The Growth of Political Stability in England: 1675–1725* (London: Macmillan, 1967).
20. Charles Tilly, "Parliamentarization of Popular Contention in Great Britain, 1758–1834," *Theory and Society* 26:2/3 (1997): 245–273; John Brewer, *Party Ideology and Popular Politics at the Accession of George III* (Cambridge: Cambridge University Press, 1976); Kathleen Wilson, "Empire, Trade and Popular Politics in Mid-Hanoverian Britain: The Case of Admiral Vernon," *Past and Present* 121 (1988): 74–109; Kathleen Wilson, *The Sense of the People: Politics, Culture and Imperialism in England, 1715–1785* (Cambridge: Cambridge University Press, 1995); Gillian Russell, *The Theatres of War: Performance, Politics, and Society, 1793–1815* (Oxford: Clarendon Press, 1995).
21. Henry Sacheverell Homer, *An Enquiry into the Means of Preserving and Improving the Publick Roads of This Kingdom* (Oxford: S. Parker, 1767); Thompson Cooper, "Homer, Henry Sacheverell (*bap.* 1719, *d.* 1791)," rev. Jeffrey Herrle, *Oxford Dictionary of National Biography* (Oxford: Oxford University Press, 2004).
22. Jessica Riskin, "The 'Spirit of System' and the Fortunes of Physiocracy," in Margaret Schabas and Neil De Marchi, eds., *Oeconomies in the Age of Newton* (Durham, N.C.: Duke University Press, 2003), 42–73; Rosalind Williams, "Cultural Origins and Environmental Implications of Large Technological Systems," *Science in Context* 6:2 (2008): 377–403.

23. Adam Smith, *An Inquiry into the Nature and Causes of the Wealth of Nations*, 2 vols. (London: W. Strahan and T. Cadell, 1776), 1:183
24. Bernard Mandeville, *The Fable of the Bees* (London: Printed by J. Tonson, 1729), 365.
25. Ibid., 263.
26. Ibid.
27. Adam Smith, *An Inquiry into the Nature and Causes of the Wealth of Nations*, 2:336.
28. Catherine Sinclair, *Sir John Sinclair, Bart.: A Memoir* (Edinburgh, 1836); John Sinclair, *The Correspondence of the Right Honourable Sir John Sinclair, Bart.* (London: H. Colburn and R. Bentley, 1831); John Sinclair, *An Account of the Highland Society of London, from Its Establishment in May 1778, to the Commencement of the Year 1813* (London, 1813); Rosalind Mitchison, *Agricultural Sir John: The Life of Sir John Sinclair of Ulbster, 1754–1835* (London: Bles, 1962).
29. John Sinclair, *Specimens of Statistical Reports* (London: T. Cadell, 1793); John Sinclair, *The Statistical Account of Scotland* (Edinburgh: William Creech, 1798); John Sinclair, *History of the Origin and Progress of The Statistical Account of Scotland* (London: Messrs Robinson, 1798); Maisie Steven, *Parish Life in Eighteenth-Century Scotland: A Review of the Old Statistical Account* (Aberdeen: Scottish Cultural Press, 1995); Maisie Steven, *Gems of Old Scotland: Scenes and Stories from the Old Statistical Account* (Glendaruel: Argyll, 2008).
30. "Advertisement," *Communications to the Board of Agriculture*, 7 vols. (London: W. Bulmer and Co., 1797–1813), 1:i. See also John Sinclair, "Address to the Board of Agriculture," July 29, 1794, quoted in Robert Beatson, "Observations on Making and Repairing Roads, Wherein Are Suggested Several Improvements on Their Construction, and on Wheel Carriages," *Communications to the Board of Agriculture*, 1:119–161; and John Sinclair, *Account of the Origin of the Board of Agriculture* (London: W. Bulmer and Co., 1796).
31. John Sinclair, "Preliminary Observations on the Origin of the Board of Agriculture," *Communications to the Board of Agriculture*, 1:iv.
32. John Wright of Chelsea, "Observations on the Public Roads of the Kingdom, and the Means of Improving Them," *Communications to the Board of Agriculture*, 1:162–165.
33. C. W. Sutton, "Holt, John (bap. 1743, d. 1801)," rev. Anita McConnell, *Oxford Dictionary of National Biography* (Oxford: Oxford University Press, 2004).
34. John Holt, *General View of the Agriculture of the County of Lancaster* (London: Printed for G. Nicol, 1795), 195.
35. "Beatson," M. F. Conolly, *Biographical Dictionary of Eminent Men of Fife* (Edinburgh: Inglis & Jack, 1866), 39; Francis Espinasse, "Beatson, Robert

(1741–1818)," rev. Philip A. Hunt, *Oxford Dictionary of National Biography* (Oxford: Oxford University Press, 2004).

36. Beatson, "Observations on Making and Repairing Roads," 1:124.
37. Robert Southey, *A Letter to Robert Smith, Esq., MP* (London: J. Murray, 1817), 33.
38. The committees initially appointed to look into the possibility of a route from London through Wales include Committee on Holyhead Roads and Harbor, *First Report,* 1810 (166); *Second Report,* 1810 (352). Committee on Holyhead Roads, *Report,* 1810–11 (197); Select Committee on Holyhead Roads, *Report,* 1814–15 (363); *Second Report,* 1814–15 (395). Select Committee on the Roads from Holyhead to London, *First Report,* 1817 (313); *Second Report,* 1817 (332); *Third Report,* 1817 (411); *Fourth Report,* 1817 (459); *Fifth Report,* 1817 (469). Select Committee on the Road from London to Holyhead, *First Report,* 1819 (78); *Second Report,* 1819 (217); *Third Report,* 1819 (256); *Fourth Report,* 1819 (501); *Fifth Report,* 1819 (549). Select Committee on the Road from London to Holyhead, *First Report (Bridge at Conway),* 1820 (201); *Second Report (Turnpike Trusts),* 1820 (224); Select Committee on the Roads from London to Holyhead, *Third Report (Chester Road),* 1822 (875); *Fourth Report (Road through England, from North Wales to London),* 1822 (343). Commissioners Appointed under the Act of 4 Geo. IV. C. 74. For Vesting in Them Certain Bridges Now Building, &c, and for the Further Improvement of the Road from London to Holyhead, *First Report,* 1824 (305); *Second Report,* 1825 (492); *Third Report,* 1826 (129); *Fourth Report,* 1826–7 (412); *Fifth Report,* 1828 (476); *Sixth Report,* 1829 (316); *Seventh Report,* 1830 (659); *Eighth Report,* 1831 (280); *Ninth Report,* 1831–2 (584); *Tenth Report,* 1833 (739); *Eleventh Report,* 1834 (608); *Twelfth Report,* 1835 (554); *Thirteenth Report,* 1836 (437); *Fourteenth Report,* 1837 (533); *Fifteenth Report,* 1837–8 (716); *Sixteenth Report,* 1839 (491); *Seventeenth Report,* 1840 (597); *Eighteenth Report,* 1841 (425); *Nineteenth Report,* 1842 (572); *Twentieth Report,* 1843 (628); *Twenty-First Report,* 1844 (277, 630); *Twenty-Second Report,* 1845 (616); *Twenty-Third Report,* 1846 (718); *Twenty-Fourth Report,* 1847 (604); *Twenty-Sixth Report,* 1849 (612); *Twenty-Seventh Report,* 1850 (753); *Twenty-Eighth Report,* 1851 (652). The technical details of the construction, the stagecoaches, and the mail route to Ireland are the subject of numerous secondary publications: Samuel Smiles, *Lives of the Engineers* (London: John Murray, 1861), esp. chap. 11, "The Holyhead Road"; Mervyn Hughes, "Telford, Parnell, and the Great Irish Road," *Journal of Transport History* 6:4 (1964): 202–204; H. C. F. Lansberry, "James McAdam and the St. Albans Turnpike Trust," *Journal of Transport History* 7:2 (1965): 121–123; and Derrick Beckett, *Telford's Britain* (London: David and Charles, 1987), esp. chap. 6, "The London and Holyhead Road." Like the works on the Highland Roads and Bridges project, these too tend to emphasize Telford's development as an engineer and his accomplishments in bridge building.

39. Colonel Tittler, *Ireland Profiting by Example* (Dublin: J. Millikin, 1799). Tittler noted with approbation the £30,000 forfeited estates grant secured for road building by the Highland Society and observed that "from the year 1770, a regular allowance has been granted by Parliament, along with the public supplies, of a sum betwixt 5000l and 9000l sterling, per annum, for the constructing and repairing of roads, and the building and support of bridges in the Highlands of Scotland." Ibid., 27.
40. "On the Origin and Progress of the Highland Society of Scotland, and the Public Advantages which have been derived therefrom," appendix 5 in Select Committee to Examine Accounts from Exchequer of Scotland on Grants Payable out of Funds of Forfeited Estates in Scotland, *Report*, 1806 (221).
41. Select Committee on State of Roads from Holyhead to London, and Conveyance of H. M. Mail between London and Dublin, *Report*, 1814–15 (363); Select Committee on the Roads from Holyhead to London, *First Report*, 1817 (313); Select Committee on State of Road from London to Holyhead, by Coventry and Shrewsbury, *Report*, 1819 (217); Select Committee on State of Roads from London and Chester to Holyhead, *Report*, 1822 (41); Select Committee on State of Public Income and Expenditure of United Kingdom, *Report, Report*, 1828 (110); *Second Report*, 1828 (420); *Third Report*, 1828 (480); *Fourth Report*, 1828 (519); Select Committee on Holyhead and Liverpool Roads, *Report*, 1830 (432).
42. Fiscal reformers from the Treasury like Nicholas Vansittart provided outside oversight. Commissioners of Highland Roads and Bridges, *Statement of Origin and Extent of Roads in Scotland, and Papers Relating to Military Roads, 1813–1814* (63), 6.
43. Select Committee on Report by T. Telford for Facilitating and Improving Communication between England and Ireland from Carlisle by N. W. of Scotland, 1809 (269).
44. Charles Arbuthnot of Rockfleet Castle, County Mayo, a secretary of the Treasury, was put in charge of the expenses for the Holyhead Road. The others were Charles Williams Wynn of Denbighshire, MP for Montgomeryshire; John Thomas Stanley, heir to Penrhos near Holyhead in Wales; T. Peers Williams of Anglesey; and Thomas Mostyn of Carnavonshire, MP for Flint. Committee for Improvement of Road from London to Holyhead, *Fifth Report*, 1828 (476), 5; *Estimate of Sum Required in 1821 for Commissioners of Holyhead Roads to Complete Improvements*, 1821 (633).
45. There were two Irishmen: Henry Parnell of Queen's County, and J. Leslie Foster of Dublin, MP for Armagh. There were four Scots: Charles Arbuthnot, chair; Robert Williams, representative for Caernarvon; J. C. Herries of Dumfriesshire, a commissioner for the Treasury; and Charles Williams Wynn of Wynnstay, Denbighshire, MP for Montgomeryshire. The others were John

Thomas Stanley, MP for Cheshire; G. Irby of Boston; Augustus Elliot Fuller of Sussex; W. L. Hughes of Berkshire; and William Smith, the Norwich banker. Commissioners for Improvement of Road from London to Holyhead, *Report*, 1824 (305).

46. John Loudon McAdam, *Remarks on the Present System of Road Making* (London: Printed for Longman, Hurst, Rees [1816] 1824); Henry Parnell, *A Treatise on Roads* (London: Longman, Rees, Orme, Brown, Green and Longman, 1833); Thomas Telford, *The Life of Thomas Telford, Civil Engineer, Containing a Descriptive Narrative of His Professional Labours* (London: Payne and Foss, 1838).

47. For example, the role played by the Post Office surveyors in justifying the controversial Highgate Archway in 1810–1812. Untitled document, Post 10/252.

48. Early committees under Browne include Commissioners of Highland Roads and Bridges, *Statement of Origin and Extent of Roads in Scotland, and Papers Relating to Military Roads*, 1813–1814 (63), 6. Hawkins chaired the following committees: Select Committee on Survey and Report of Coasts and Central Highlands of Scotland, *First Report*, 1802–3 (80); *Second Report*, 1802–3 (94), *Third Report*, 1802–3 (110), and *Fourth Report*, 1802–1803 (118). Sinclair chaired the following committees: Select Committee on Acts Regarding Use of Broad Wheels, *First Report*, 1806 (212), and *Second Report*, 1806 (221); *First Report*, 1808 (76); *Second Report*, 1808 (77); *Report*, 1809 (238); *Second Report*, 1809 (238); *Third Report*, 1809 (271); Committee on the Highways of the Kingdom, *First Report*, 1808 (225); *Second Report*, 1808 (275); *Third Report*, 1808 (315); Select Committee on Acts Regarding Turnpike Roads and Highways in England and Wales, *Report*, 1810–1811 (240). Sir James Graham chaired the following committees: Select Committee on Report by T. Telford for Facilitating and Improving Communication between England and Ireland from Carlisle by N. W. of Scotland, 1809 (269); and Select Committee on Reports on Line of Roads from Dumfries to Newton Stewart, and Communication between England and Ireland from Carlisle by N. W. Scotland, *Report*, 1810–1811 (119).

49. Ian Newbould, *Whiggery and Reform, 1830–41: The Politics of Government* (Houndsmills, Basingstoke, Hampshire: Macmillan, 1990).

50. Smith, *Inquiry into the Nature and Causes of the Wealth of Nations*, 2:339, quoted in Beatson, "Observations on Making and Repairing Roads," 1:123, 127.

51. Select Committee on Acts regarding Turnpike Roads and Highways in England and Wales, *Report*, 1810–1811 (240), 5.

52. Select Committee on Report, Plan and Estimate of Road from Carlisle to Glasgow, *Report*, 1814–1815 (463), 5–6.

53. Smith, *Inquiry into the Nature and Causes of the Wealth of Nations*, 1:184.

54. Homer, *Enquiry into the Means of Preserving and Improving the Publick Roads of This Kingdom*, 22.

55. Simon Morris, "The Marylebone and Finchley Turnpike, 1820–50," *Camden History Review* 21 (1998): 24–32.
56. Beatson, "Observations on Making and Repairing Roads," 1:121.
57. Thomas Grahame, *A Treatise on Internal Intercourse and Communication in Civilized States, and particularly in Great Britain*, 8 vols. (London: Longman, Rees, Orme, Brown, Green, and Longman, 1834), 1:13–14.
58. Select Committee on Acts Regarding Use of Broad Wheels, *Second Report*, 1806 (221), 51; *First Report*, 1808 (225), 79, 107
59. Select Committee on State of Road from London to Holyhead, by Coventry and Shrewsbury, *Sixth Report*, 1819 (549), 102.
60. Select Committee on Acts Regarding Use of Broad Wheels, *First Report*, 1808 (225), 79, 107; *Second Report*, 1806 (221), 51.
61. John Holt, *General View of the Agriculture of the County of Lancaster* (London: G. Nicol, 1795), 196.
62. Beatson, "Observations on Making and Repairing Roads," 1:121.
63. Grahame, *Treatise on Internal Intercourse and Communication*, 1:13.
64. Holt, *General View*, 195.
65. "Metropolis Turnpike Trusts," *Parliamentary Debates* 4 (February 27, 1821): 946–947.
66. Wright, "Observations on the Public Roads of the Kingdom," 1:166.
67. Select Committee on Acts Regarding Use of Broad Wheels, *First Report*, 1808 (225), 7, 107; *Second Report*, 1806 (221), 191.
68. Parnell, *Treatise on Roads*, 293.
69. Beatson, "Observations on Making and Repairing Roads," 1:121.
70. John Sinclair, *Statistical Account of Scotland*, 1:xii.
71. Ibid., xxxiv.
72. "Advertisement," *Communications to the Board of Agriculture*, 1:i. See also John Sinclair, *Plan for Establishing a Board of Agriculture and Other Internal Improvements*.
73. They included Richard Watson, bishop of Llandaff; Thomas Robertson, a Church of Scotland minister from Lauder and Dalmeny who penned the report on Dalmeny; Bryce Johnstone, a Church of Scotland minister from Annan who penned the Dumfries report; George Skene Keith, a Church of Scotland minister from Mar, near Aberdeen (Aberdeenshire); Robert Douglas, a Church of Scotland minister from Perthshire (Roxburgh and Selkirk); and John Fleming, a Free Church of Scotland minister from Lenlithgow (Shetland). Other correspondents were Scottish land surveyors, farmers, and mathematicians, some only Scottish by schooling or marriage: John Shirreff of Haddingtonshire; George Robertson of Edinburghshire (Midlothian; Kincardineshire); George Steuart Mackenzie of Edinburgh (Ross and Cromarty); David Ure of Glasgow (Dunbartonshire, Roxburghshire, Kinrossshire); James Anderson of

Hermiston, near Edinburgh, a farmer (Aberdeenshire); Thomas Bates of Northumberland, a graduate of the University of Edinburgh; James Donaldson of Dundee (Banff, Elgin, Nairn, Kincardie, Northampton, and Perth); and John Fuller of Berwickshire (Berwick). But even many of the reports from English counties were authored by Scots. George Rennie of Haddingtonshire wrote the report on the West Riding of Yorkshire and William Fordyce Mavor of Aberdeenshire wrote that on Berkshire.

74. Beatson, "Observations on Making and Repairing Roads," 1:122.
75. John Holt of Walton, "Hints on the Subject of Roads," *Communications to the Board of Agriculture*, 1:183.
76. Beatson, "Observations on Making and Repairing Roads," *Communications to the Board of Agriculture*, 1:128–36.
77. Ibid., 1:150, 156.
78. John Sinclair, *Sketch of an Introduction to the Proposed Analysis of the Statistical Account of Scotland* (London: Printed by W. Bulmer and Co, 1802), 15.
79. Ibid., 16.
80. Rosalind Mitchison, "Sinclair, Sir John, first baronet (1754–1835)," *Oxford Dictionary of National Biography* (Oxford: Oxford University Press, 2004).
81. *Public Characters of 1798–9* (London: R. Phillips, 1799), 60.
82. Ibid., 59–64.
83. Beatson, "Observations on Making and Repairing Roads," 1:124.
84. Select Committee on Acts Regarding Use of Broad Wheels, and Regulations for Preservation of Turnpike Roads and Highways of United Kingdom, *First Report*, 1806 (212); Select Committee on Acts Regarding Turnpike Roads and Highways in England and Wales, *Report*, 1810–1811 (240).
85. Select Committee on Acts Regarding Use of Broad Wheels, *Third Report*, 1809 (271), 46.
86. "Interview, Thomas Telford," Select Committee on the Highways of the Kingdom, *Report*, 1819 (509), 54–57; McAdam, *Remarks on the Present System of Road Making* (note 44).
87. Thomas Telford, "A Survey and Report of the Coasts and Central Highlands of Scotland," *Parliamentary Papers* 1802–03 (45), 3.
88. Select Committee on State of Road from London to Holyhead, by Coventry and Shrewsbury, *Sixth Report*, 1819 (549), 100.
89. Thomas Telford, "A Survey and Report of the Coasts and Central Highlands of Scotland," *Parliamentary Papers* 1802–03 (45), 3.
90. Select Committee on Acts Regarding Use of Broad Wheels, and Regulations for Preservation of Turnpike Roads and Highways of United Kingdom, *First Report*, 1806 (212), 51; *Second Report*, 1808 (77), 79, 107.
91. Committee on the Highways of the Kingdom, *First Report*, 1808 (225), 7.
92. Beatson, "Observations on Making and Repairing Roads," 1:121.

93. Select Committee on Acts Regarding Use of Broad Wheels, *Second Report*, 1806, 51; *Second Report*, 1808 (77), 79, 107
94. Parnell, *Treatise on Roads*, 292–293.
95. Select Committee on Acts Regarding Use of Broad Wheels, and Regulations for Preservation of Turnpike Roads and Highways of United Kingdom, *First Report*, 1806 (212), 51; *Second Report*, 1808 (77), 79, 107.
96. Parnell, *Treatise on Roads*, 293–294.
97. Scott, *Digests of the General Highway and Turnpike Laws*, advertisement bound with the book, i–ii.
98. In 1797, John Wright of Chelsea looked back at the striking contrast between avid attempts at regulation under George II and the almost total absence of regulation since 1766. Wright, "Observations on the Public Roads of the Kingdom."
99. Ibid., 171.
100. Catherine Sinclair, *Shetland and the Shetlanders; Or, The Northern Circuit, Dedicated to the Highland Society* (New York: D. A. Appleton and Co., 1840), 23–24.
101. Smith, *Wealth of Nations*, 1:183.
102. John Sinclair, *A History of the Public Revenue of the British Empire*, 2 vols. (London: Printed by A. Strahan for T. Cadell and W. Davies, [1785] 1803), 2:383–384.
103. Beatson, "Observations on Making and Repairing Roads," 119.
104. Holt, *General View*, 195
105. Catherine Sinclair, *Shetland and the Shetlanders*, 23.
106. Highland Society of London, *Rules of the Highland Society of London* (London, 1783).
107. *Scottish Song in Two Volumes* (London, 1794), 1:cxviii.
108. "On the Origin and Progress of the Highland Society of Scotland," appendix 5 in Select Committee to Examine Accounts, *Report*, 1806 (221); Member of the Highland Society in London, *The Necessity of Founding Villages Contiguous to Harbours* (London: C. Macrae, 1786); George Pitcairn, *A Retrospective View of the Scots Fisheries* (Edinburgh, 1787); P. White, *Observations upon the Present State of the Scotch Fisheries* (Edinburgh: Grant and Moir, 1791); John Lanne Buchanan, *A General View of the Fishery of Great Britain* (London: T. N. Longman, 1794); Highland Society of Scotland, *Report of the Committee* (Edinburgh: W. Creech, 1790).
109. Maxine Berg, *The Age of Manufactures, 1700–1820: Industry, Innovation, and Work in Britain* (London: Routledge, [1985] 1994).
110. David Pearce, *London—Capital City* (London: Batsford, 1988).
111. Patrick Woodland, "Political Atomization and Regional Interests in the 1761 Parliament: The Impact of the Cider Debates, 1763–1766," *Parliamentary*

History 8:1 (1989): 63–89; Brewer, *Party Ideology and Popular Politics at the Accession of George III*.

112. Commissioners of Highland Roads and Bridges, *Statement of Origin and Extent of Roads in Scotland, and Papers Relating to Military Roads*, 1813–1814 (63), 19.
113. Select Committee on Report, Plan and Estimate of Road from Carlisle to Glasgow, *Report*, 1814–1815 (463), 9.
114. Ibid., 5–6.
115. *Parliamentary Register* 4 (June 15, 1803): 171.
116. Select Committee on Broad Wheels, *Second Report*, 1808 (77), 184.
117. John Sinclair, *History of the Public Revenue of the British Empire*, 2:383–384.
118. Commissioners of Highland Roads and Bridges, *Statement of Origin and Extent of Roads in Scotland, and Papers Relating to Military Roads*, 1813–1814 (63), 55.
119. Select Committee on Acts Regarding Turnpike Roads and Highways in England and Wales, *Report*, 1810–1811 (240), 5.
120. Select Committee on Acts Regarding Use of Broad Wheels, *First Report*, 1808 (225), 6
121. Thomas Telford, "A Survey and Report of the Coasts and Central Highlands of Scotland," *Parliamentary Papers* 1802–03 (45), 23.
122. Holt, *General View*, 195.
123. Ibid.
124. Ibid., 23.
125. John Sinclair, *History of the Public Revenue of the British Empire*, 2:383–384.
126. George Dempster, *A Discourse Containing a Summary of the Proceedings of the Directors of the Society for Extending the Fisheries* (London: G. and T. Wilkie, 1789), 11.
127. *Parliamentary Register*, 171.
128. Thomas Telford, "A Survey and Report of the Coasts and Central Highlands of Scotland," *Parliamentary Papers* 1802–03 (45), 3.
129. Thomas Telford, "On the Emigrations from the Highlands," *Scots Magazine* (1803): 329. Emigration from Scotland rose markedly between 1801 and 1803, and the Highland Society began submitting reports to the government containing the details of this trend. The "Report from the Highland Society of Scotland" presented as an appendix to Telford's 1803 report, "On the Emigrations from the Highlands," asked its correspondents to answer the question, "Would the undertaking these Public Works at the present Time, by affording Employment to the People, giving them Habits of Industry, and furnishing them with Capital, tend to check the Spirit of Emigration which now prevails." See also "On the Origin and Progress of the Highland Society of Scotland, and the Public Advantages Which Have Been Derived Therefrom," appendix 5 in

Select Committee to Examine Accounts from Exchequer of Scotland on Grants Payable out of Funds of Forfeited Estates in Scotland, *Report*, 1806 (221).
130. *Parliamentary Register*, 171.
131. "Agricultural Distress," *Hansard's Parliamentary Debates*, vol. 33 (March 22, 1816), 530.
132. "Extract of a Letter from the Rev. John Proctor of Ippolitz near Hitchin, Herts," Select Committee on Acts Regarding Use of Broad Wheels, *Third Report*, 1808 (315), 9.
133. Committee on Holyhead Roads and Harbor, *First Report*, 1810 (166); *Second Report*, 1810 (352); Committee on Holyhead Roads, *Report*, 1810–11 (197); Select Committee on Holyhead Roads, *Report*, 1814–15 (363); *Second Report*, 1814–15 (395).
134. Thomas Telford, "A Survey and Report of the Coasts and Central Highlands of Scotland," *Parliamentary Papers* 1802–03 (45), 4.
135. Select Committee on Report by T. Telford for Facilitating and Improving Communication between England and Ireland from Carlisle by N. W. of Scotland, 1809 (269), 29.
136. "Instances of Stage Coaches Recently Broken Down or Overturned," an interview with F. Dickins in Wollaston, Northamptonshire, by John Sinclair, in 1808, appendix in Select Committee on Acts Regarding Use of Broad Wheels, *Third Report*, 1808 (315), 203. For the use of similar arguments on a local level, see Committee on Repealing the Act of 1810, "For Repairing and Maintaining the Road from the Eynesford Turnpike Road," *Minutes of Evidence*, May 30, 1811.
137. "Extract of a Letter from the Rev. John Proctor of Ippolitz near Hitchin, Herts."
138. Select Committee on Acts Regarding Use of Broad Wheels, *First Report*, 1806 (212), 6.
139. Ibid. See also "Report from the Committee on Petitions Relating to the General Post-Office," Post Office Archive 91/2, July 26, 1814.
140. Select Committee on Acts Regarding Use of Broad Wheels, *First Report*, 1806 (212), 4.
141. Select Committee on Report, Plan and Estimate of Road from Carlisle to Glasgow, *Report*, 1814–1815 (463), 3.
142. Select Committee on Acts Regarding Use of Broad Wheels, *First Report*, 1806 (212). 6.
143. *Annual Register*, 1814, 8–10; William Hone, *The Every-Day Book*, 2 vols. (London: Published for W. Hone by Hunt and Clarke, 1826), 2:101–114.
144. "Severity of the Weather," *Times*, January 26, 1814, 2d; January 24, 3e; January 26, 2d.
145. "Pavement and Roads VII," *Times*, June 3, 1816, 3d; "Pavement and Roads III: Present Management of the Pavements," *Times*, January 2, 1816, 2b; "House of

Commons, Thursday, March 7: Pavement of the Metropolis," *Times*, March 8, 1816, 2b.
146. The first British image to address the imperiled stagecoach is Philip James de Loutherbourg's *Landscape with Carriage in a Storm* (1804). James Pollard also produced a number of images depicting specific coaches lost in the snow. See *The York and Edinburgh Royal Mail Coach Stuck in a Snowdrift* (1821) and *The Birmingham-London Royal Mail*, as well as Charles Cooper Henderson's set of four *Mail Coach Incidents* of 1820–1830. Images are held at the study library of the Yale Center for British Art.
147. See also *The Coachman Walking at the Head of His Team*, the series *Stage Coach Snow-bound*, and *Devonshire Mail near Amesbury in an Avalanche*, all by Henry Alken.
148. John Sinclair, *General Report of the Agricultural State, and Political Circumstances, of Scotland* (Edinburgh: Abernethy and Walker, 1812), 443.
149. Robert Southey, *Letters from England* (New York: David Longworth, 1803), 56.
150. Select Committee on the Roads from Holyhead to London, *First Report*, 1817 (313), 9.
151. Thomas Hughes, *The Practice of Making and Repairing Roads* (London: J. Weale, 1838), 11.
152. *The Roads and Railroads, Vehicles, and Modes of Travelling, of Ancient and Modern Countries* (London: John W. Parker, 1839), 80.
153. Parnell, *Treatise on Roads*, 132.
154. Thomas Hughes, *Practice of Making and Repairing Roads*, 14.
155. McAdam, *Remarks on the Present System of Road Making*, 148.
156. Thomas Hughes, *Practice of Making and Repairing Roads*, 14.
157. T. Telford, "Report on State of Road between London and Shrewsbury," Commissioners for Improvement of Holyhead Roads, *Report*, 1820 (126), 12.
158. McAdam, *Remarks on the Present System of Road Making*, 9, 149.
159. McAdam, *Remarks on the Present System of Road Making*; Parnell, *Treatise on Roads*; Richard Lovell Edgeworth, *An Essay on the Construction of Roads and Carriages* (London: Printed for J. Johnson, 1813); Edmund Leahy, *A Practical Treatise on Making and Repairing Roads* (London: J. Weale, 1844); Henry Law and Samuel Hughes, *Rudiments of the Art of Constructing and Repairing Common Roads* (London: J. Weale, 1850).
160. Quoted in Simeon DeWitt Bloodgood, *A Treatise on Roads, Their History, Character, and Utility* (Albany: O. Steele, 1836), 75–76.
161. Select Committee on State of Road from London to Holyhead, by Coventry and Shrewsbury, *Sixth Report*, 1819 (549), 102.
162. Bill to Amend Laws Regarding Turnpike Roads and Highways, 1809 (279); 1809 (102); 1809 (201); 1809 (255(20)); 1809 (347); 1810–1811 (48); 1810–1811 (93).
163. *Parliamentary Debates* 34 (May 2, 1820): 53.

164. Charles Dupin, *The Commercial Power of Great Britain*, 2 vols. (London: C. Knight, 1825), 1:50.
165. 5 & 6 Will. 4, c. 50, sect. 56 (General Highway Act, 1835).
166. The project was first discussed within the purview of a Highland Society project for restoring Britain's fisheries, on which Sinclair was also active during the 1790s, and which in 1801 commissioned Telford to survey the whole of Scotland. The fisheries committee was then reconstructed as the Select Committee on the Survey and Report of the Coasts and Central Highlands of Scotland. Its *Second Report*, in 1802–1803, dealt precisely with the problem of the highways and laid the groundwork for the Commission on Roads and Bridges in Scotland. See Select Committee on Survey and Report of Coasts and Central Highlands of Scotland, *Second Report*, 1802–3 (94); Commissioners of Highland Roads and Bridges, *First Report*, 1803–1804 (108); *Second Report*, 1805 (176).
167. Commissioners of Highland Roads and Bridges, *Statement of Origin and Extent of Roads in Scotland, and Papers Relating to Military Roads*, 1813–1814 (63), 6.
168. After new roads were built, a new act, 50 Geo. 3, c. 43 (1810), put maintenance in the hands of the counties and the heritors of adjoining lands.
169. Commissioners of Highland Roads and Bridges, *Statement of Origin and Extent of Roads in Scotland, and Papers Relating to Military Roads*, 1813–1814 (63), 8.
170. In the 1810 bill, the committee refused to pay, "which may be looked upon as an experiment whether the Highland Proprietors would exert themselves to maintain in repair Roads towards which they had then contributed about Ninetey Thousand Pounds"; the grim answer: "The experiment has entirely failed, and the Roads first finished now evidently falling to decay, it has appeared to be the duty of the Commissioners appointed by The Highland Road and Bridge Act, to collect information, and to give attentive consideration to the subject in all its parts." Ibid., 8.
171. Ibid., 3–8.
172. Ibid., 9.
173. "The Life and Labours of Telford," *Mechanics' Magazine* 872 (April 25, 1840): 653; actually, only £365,000 of this was the parliamentary contribution for roads themselves in the Highlands. The remainder was for bridges, harbors, and lowland roads that form part of the total. An additional £550,000 was paid for by local contributions. Commissioners of Inquiry into the State of the Law and Practice in Respect to the Occupation of Land in Ireland, *Digest of Evidence*, 1845, 605.
174. There were 3,666 miles of turnpike roads in Scotland in 1829, managed by 190 turnpike trusts. J. R. McCulloch, *A Statistical Account of the British Empire*, 2 vols. (London: Charles Knight and Co., 1839), 2:48.

175. "Life and Labours of Telford," 653.
176. Ibid., 564.
177. 55 Geo. 3, c. 152.
178. Dupin, *Commercial Power of Great Britain*, i:85–86; Select Committee on State of Road from London to Holyhead, by Coventry and Shrewsbury, *Sixth Report*, 1819 (549).
179. Select Committee on State of Road from London to Holyhead, by Coventry and Shrewsbury, *First Report*, 1817 (313), 188–189.
180. Select Committee on State of Road from London to Holyhead, by Coventry and Shrewsbury, *Sixth Report*, 1819 (549), 104.
181. 59 Geo. 3, c. 30; Telford, "Annual Report on Holyhead Road," Commissioners Appointed under the Act of 4 Geo. IV. c. 74., *First Report*, 1824 (305), 17ff.
182. Committee on Holyhead Roads and Harbor, Second *Report*, 1810 (352). 7, 15, 18.
183. Thomas Telford, "Report . . . to the Commissioners Acting under the 55 Geo. III, for the Improvement of the Holyhead Roads," in Select Committee on State of Roads from London and Chester to Holyhead, *First Report (N. Wales)*, 1822 (41), 5.
184. Select Committee on Holyhead and Liverpool Roads, *Report*, 1830 (432), 5.
185. Smiles, *Lives of the Engineers*, 256.
186. Telford, "Report on the Road in North Wales," in Select Committee on State of Road from London to Holyhead, by Coventry and Shrewsbury, *Sixth Report*, 1819 (549), 104.
187. "Comparative Statement of the Travelling through One of the Gates in the Commissioners' Hands," Commissioners of Shrewsbury and Holyhead Roads, *Report*, 1828 (168), 4.
188. 55 Geo. 3, c. 152; Telford, "Report on the English Part of the Holyhead Road," in Select Committee on State of Road from London to Holyhead, by Coventry and Shrewsbury, *Sixth Report*, 1819 (549), 105–106.; T. Telford, "Reports on State of Road between London and Shrewsbury," in Commissioners for Improvement of Holyhead Roads, *Report*, 1820 (126); *Estimate of Sum for Completing Line of Road between Chirk and Bangor Ferry in N. Wales*, 1820 (194), 182.
189. Select Committee on State of Road from London to Holyhead, *Second Report (Turnpike Trusts)*, 1820 (224), 224.
190. 1 Geo. 4, c. 70; 2 Geo. 4, c. 74. *Estimate of Sum Required in 1821 for Commissioners of Holyhead Roads to complete Improvements*, 1821 (633), 1. Bridges were put under their authority by 4 Geo. 4, c. 74, but this act expired July 11, 1825.
191. Commissioners of Holyhead Roads, Holyhead and Howth Harbours, and Menai and Conway Bridges, *Account of Sums Advanced*, 1826 (355).
192. 6 Geo. 4, c. 100; Commissioners for Improvement of Road from London to Holyhead, and from London to Liverpool, *Fourth Report*, 1826–1827 (412).

193. Select Committee on State of Road from London to Holyhead, *Second Report (Turnpike Trusts)*, 1820 (224), 12.
194. Select Committee on Holyhead and Liverpool Roads, *Report*, 1830 (432).
195. T. Telford, "Report on State of Road from London to Liverpool," 1826–1827 (362); 7 & 8 Geo. 4, c. 35, "An Act for the Further Improvement of the Road from London to Holyhead, and of the Road from London to Liverpool," 1829 (123); T. Telford, "Reports, Estimates, and Plans for Improving Road from London to Liverpool," *House of Commons Sessional Papers, Reports of Commissioners* 1826 (362).
196. McCulloch, *Statistical Account of the British Empire*, 2:44.
197. Hermann F. H. Pückler-Muskau, *A Tour in England, Ireland and France in the years 1828 and 1829*, 2 vols. (London: Effingham Wilson, 1832), 1:16–17; *Civil Engineer and Architect's Journal*, February 1842, 58; Thomas Roscoe, *Excursions in North Wales* (London: C. Tilt and Simking and Co., 1836), 112.
198. Parnell, *Treatise on Roads*, 368–369.
199. Select Committee on State of Road from London to Holyhead, by Coventry and Shrewsbury, *Sixth Report*, 1819 (549), 99.
200. "An Account of All Sums Received and Expended by the Commissioners for the Improvement of the Road from London to Holyhead, &c., from the Institution of the Commission in 1815 to the 5th April 1830," appendix 1 in Select Committee on Holyhead and Liverpool Roads, *Report*, 1830 (432), 31; "Estimates, Miscellaneous Services," *House of Commons Sessional Papers, Reports of Commissioners*, 1823 (429); "Account of Sums Outstanding of Advances for Public Works," *House of Commons Sessional Papers, Reports of Commissioners*, 1826–1827 (405); "Account of Sums Paid into Exchequer by Post Office and Holyhead Road Commissioners for Repayment of Loans for Building of Menai and Conway Bridges," *House of Commons Papers*, 1826–1827 (292).
201. Samuel H. Williamson, "Five Ways to Compute the Relative Value of a U.S. Dollar Amount, 1790–2005," http://MeasuringWorth.Com, 2006. Williamson recommends four different indicators—price index, value of household bundle, income, and GDP—for different estimates of different kinds of projects. For a nation, collective project, measuring the cost in terms of the historically equivalent GDP, indicates correctly the imagined burden of these infrastructural projects in terms of their equivalents today.
202. *Estimate of Sum Required in 1821 for Commissioners of Holyhead Roads to Complete Improvements*, 1821 (633); "Account of Sums Paid into Exchequer by Post Office and Holyhead Road Commissioners for Repayment of Loans for Building of Menai and Conway Bridges," *House of Commons Sessional Papers*, 1826–1827 (292).
203. Select Committee on Holyhead and Liverpool Roads, *Report*, 1830 (432), 3.

204. *Estimate of Sum Required in 1821 for Commissioners of Holyhead Roads to complete Improvements,* 1821 (633); "Estimates, Miscellaneous Services," *House of Commons Sessional Papers, Reports of Commissioners,* 1823 (429); "Estimates, Miscellaneous Services (Holyhead and Howth Roads and Harbours)," *House of Commons Sessional Papers, Reports of Commissioners,* 1825 (151).
205. "Account of Sums Outstanding of Advances for Public Works," *House of Commons Sessional Papers, Reports of Commissioners,* 1826–1827 (405), 1. Other debtors for improvement to the Treasury included the Commissioners for the Improvement of the Metropolis, the Docks at Leith, and the Rebuilding of London Bridge.
206. 3 Geo. 4, c. 120 (Turnpike Act, 1822); 9 Geo. 4, c. 1 (General Turnpike Act, 1828); "An Account of Income and Expenditure of Trustees of Turnpike Trustees in England and Wales under Act 3 George IV., Chapter 120," *House of Commons Sessional Papers, Reports of Commissioners,* 1824 (470).
207. 5 & 6 Will. 4, c. 50 (General Highway Act, 1835).

3. Paying to Walk

1. David J. V. Jones, *Rebecca's Children* (Oxford: Clarendon Press, 1989); Pat Molloy, *And They Blessed Rebecca: An Account of the Welsh Toll-Gate Riots, 1839–1844* (Llandysul: Gomer Press, 1983); David Williams, *The Rebecca Riots: A Study in Agrarian Discontent* (Cardiff: University of Wales Press, 1955).
2. William Smith O'Brien, *Principles of Government; or, Meditations in Exile* (Boston, Mass.: P. Donahoe, 1856), 223–224.
3. Civil Engineer, *Personal Recollections of English Engineers* (London: Hodder and Stoughton, 1868), 424.
4. Charles Knight, *The Popular History of England* (London: Bradbury and Evans, 1862), 506.
5. Robert Kemp Philp in his *History of Progress* (London: Houlston and Wright, 1857) gave a chronology of turnpike riots concluding with Rebecca and voiced his hope that they were the "final instances of a mistaken populace rising against works of improvement." He inveighed against the "humblest classes in the state," the "first to find the benefit" of infrastructure in real economic terms, who nonetheless were also the first to rise "against works of improvement" with actual acts of violence. Robert Kemp Philp, *The History of Progress in Great Britain* (London: Houlston and Wright, [1857] 1859), 157.
6. "Rebecca Rioters in Radnorshire," *Times,* October 3, 1843, 5c.
7. For the *Times,* the real story was whether the riots were related to genuine extortion on the part of Welsh turnpike trusts, and figures and testimony about the turnpike accounts were published in extensive detail. "Rebecca Rioters," *Times,* October 2, 1843, 6a. The reporters were particularly fascinated by the

response of rioters to representatives of a closed bureaucracy, interviewing, for instance, tollgate keepers who had had bits of tollgate thrown at them. "Rebecca in Gower," *Times*, July 27, 1843, 6d (the case of turnpike bits thrown at a tollgate keeper); "Rebecca Rioters at Newbridge," *Times*, November 7, 1843, 3f (the case of a toll collector having to promise that he would collect no more tolls).

8. March witnessed the breaking of a new tollgate in Somerset. The *Times* responded with praise, noting the unjust burden the toll would have placed on the poorest members of the community. "Rebecca's Daughters in Somerset," *Times*, March 28, 1843, 6d.

9. "Rebecca in Winchester," *Times*, August 23, 1843, 5c.

10. The evidence focuses primarily on London, where the movement against turnpikes was on the rise. In September, a Mr. Hill, the porter of London University College, found a note signed "Rebecca and her Daughters" affixed to the college gate on Gower Street, demanding its removal. In the morning, the bar had disappeared. The *Times* was confused about whether the note was merely a college prank or a more serious threat, but it took the opportunity to emphasize what a public nuisance the tollgate had been to circulation around Holborn, and what a service Rebecca had done on behalf of London. "Rebecca and Her Daughters in London," *Times*, September 30, 1843, 5e. From then on, news of any rise in tolls on public roads around London was published under headings like "Rebecca Wanted in Gloucestershire," a last threat of the people's justice on behalf of public circulation. "Rebecca Wanted in Gloucestershire," *Times*, June 17, 1846, 7f.

11. "A Word on the Roads," *Chambers's Edinburgh Journal* 68 (Saturday, April 19, 1845): 242.

12. William James Linton, *Threescore and Ten Years, 1820–1890: Recollections by W. J. Linton* (New York: C. Scribner's Sons, 1894), 89–91. Writing in 1845, William Pagan, author of a treatise on road reform, called Rebecca the visible face of the "spirit of dissatisfaction" against turnpike tolls, manifested by "multitudes of people." William Pagan, *Road Reform: A Plan for Abolishing Turnpike Tolls and Statute Labour* (Edinburgh: Blackwood and Sons, 1845), 200. In 1857, *Punch* exhorted its readers to "enroll as a Rebecca, or in other words join the Toll-Reform-Association, which is pledged to present us with the freedom of the country." Mark Searle, *Turnpikes and Toll-bars* (London: Hutchinson and Co., 1930), 669. In 1863, the *British Farmer's Magazine* wished that Rebecca "had extended her campaign some way east of the Severn." "The Turnpike Gate," *British Farmer's Magazine* 44 (1863): 401.

13. Harriet Martineau, *History of the Peace: Pictorial History of England during the Thirty Years Peace, 1816–1846* (London: W. and R. Chambers, 1858), 636.

14. Miles Ogborn, "Local Power and State Regulation in Nineteenth Century Britain," *Transactions of the Institute of British Geographers* 17:2 (1992): 215–226;

Philip Harling and Peter Mandler, "From 'Fiscal-Military' State to Laissez-Faire State, 1760–1850," *Journal of British Studies* 32:1 (1993): 44–70; David Eastwood, *Government and Community in the English Provinces, 1700–1870* (New York: St. Martin's Press, 1997); Philip Harling, "Parliament, the State, and 'Old Corruption,'" in Arthur Burns and Joanna Innes, eds., *Rethinking the Age of Reform: Britain, 1780–1850* (Cambridge: Cambridge University Press, 2003), 98–113; K. D. M. Snell, *Parish and Belonging: Community, Identity, and Welfare in England and Wales, 1700–1950* (Cambridge: Cambridge University Press, 2006).

15. William Albert, "Popular Opposition to Turnpike Trusts in Early Eighteenth-Century England," *Journal of Transport History* New Series 5:1 (February 1979): 1–17; E. P. Thompson, *Whigs and Hunters: The Origin of the Black Act* (London: Allan Lane, 1975), 22, 220–227, 250–257.

16. Sidney Webb and Beatrice Webb, *English Local Government: The Story of the King's Highway* (London: Longmans, Green, 1913), esp. chap. 6; Eric Pawson, "Popular Opposition to Turnpike Trusts?" *Journal of Transport History* 5 (1984): 57–65.

17. Searle, *Turnpikes and Toll-bars*, 156, 163; John Entick, *A New and Accurate Survey and History of London* (London: Dill, 1766). After 1825, those going to church or to funerals were also exempted.

18. William Albert, *The Turnpike Road System in England, 1663–1840* (Cambridge: Cambridge University Press, 1972), 82–83.

19. William Holloway, *The History and Antiquities of the Ancient Town and Port of Rye* (London: J. R. Toulmin Smith, 1847), 458.

20. Ibid., 458, 460.

21. The diarist Richard Townley noted approvingly how those taxes obviated "the necessity of turnpike tolls, and all other impositions; so that persons may travel safely, pleasantly, and expeditiously too, without paying so *grievously* for those qualifications, as we do, in almost every part of England." Richard Townley, *A Journal Kept in the Isle of Man* (Whitehaven: T. Cadell, 1791), 18.

22. *The Statistical Account of Buteshire* (Edinburgh: W. Blackwood, 1841), 112.

23. *Penny Gazetteer*, October 8 and December 9, 1735, quoted in Thomas Wright, *Caricature History of the Georges; or, Annals of the House of Hanover* (London: Chatto and Windus, 1904), 111.

24. Philp, *History of Progress in Great Britain*, 157.

25. William Albert, "Popular Opposition to Turnpike Trusts and Its Significance," *Journal of Transport History* 5 (1984): 66–68. Albert's larger claims about an antiturnpike movement were countered by Eric Pawson, "Popular Opposition to Turnpike Trusts?" *Journal of Transport History* 5 (1984): 57–65. Pawson proved that these riots did not stem the number of turnpikes proposed and passed: the consequences of eighteenth-century protest were felt at the

level of local reform, not parliamentary policy. For further instances of the impact of riots on discussions of turnpike management, see Webb and Webb, *English Local Government*, 123–124; Albert, *The Turnpike Road System*, 123–124; and Michael Freeman, "Popular Attitudes to Turnpikes in Early-Century England," *Journal of Historical Geography* 19:1 (January 1993): 33–47.

26. One example is the long process of foot-toll abolition in Sunderland, which began in 1796 at the urging of a few popular aldermen. The corporation abolished all tolls in its district in 1851. William Fordyce, *The History and Antiquities of the County Palatine of Durham* (Newcastle: A. Fullarton and Co., 1857), 483.

27. Milnes reasoned, "This is a great injustice to me and others." Between Leven and Kirkeldy, he found seven gates on a road only 7¾ miles long. Richard Milnes, *The Warning Voice of a Hermit Abroad* (Wakefield: E. Waller, 1825), 22–23.

28. "Word on the Roads," 242. The village of Cupar in Fife was ringed by "thirteen toll-bars within a circuit of three miles of the market cross." Pagan, *Road Reform*, 205. In 1857, William O'Brien described how turnpike gates still proved an unfair burden: "A person who lives within a hundred yards of a turnpike gate pays as much on passing through the gate as if he had traveled ten miles of the road." O'Brien, *Principles of Government*, 223–224.

29. *Cobbett's Two-Penny Trash; or, Politics for the Poor*, 1830, 18.

30. Milnes reckoned that the cost of turnpikes fell hardest on "poor carriers" and recounted the case of a miller who paid £6 because the heavy boiler he transported past Wakefield was carried "upon six or nine inch wheels." He condemned acts of 1822 and 1823 that raised new tolls from the one-horse carts typical of poorer tradesmen and exempted "carts for light goods and passengers." He cursed the turnpikes, the toll-bar farm, and the weighing machines in particular as "an useless needless plague to the public, and of no use to the road." Milnes reckoned that turnpikes and toll-bar farmers "cost the nation ten millions per annum, more or less," most of it visited on the heads of the poor, "more than any other tax," to the tune of "four millions per annum." Milnes, *Warning Voice of a Hermit Abroad*, 21–24.

31. For these poor persons to participate in commerce, they could not be charged the same toll as horses, lest the charge "amount to more money than their owners could earn." *Cobbett's Two-Penny Trash*, 1830, 18

32. Turnpike gates acted as a deterrence to the principle of circulation "by preventing little excursions, which would otherwise be made for the sake of health or of enjoyment." O'Brien counted these checks on participation, learning, and health as "a very great amount of privation . . . imposed upon the community." O'Brien, *Principles of Government*, 223–224.

33. *Communications to the Board of Agriculture*, 7 vols. (London: W. Bulmer and Co., 1797–1813), 1:5, 216. Recalling radical concerns, Charles Penfold, author of a utopian treatise on the centralization of highways, testified before the 1836 committee about the necessity of abolishing tolls throughout the realm, and in particular around "any large city or town." Interview, Charles Penfold, May 9, 1836, in Select Committee on Turnpike Trusts and Tolls, *Report*, 1836 (547), 56. See also Select Committee on Acts Regarding Use of Broad Wheels, *First Report*, 1806 (212), appendix 6.
34. See also Patrick Carroll-Burke, "Material Designs: Engineering Cultures and Engineering States—Ireland, 1650–1900," *Theory and Society* 31 (2002): 75–114.
35. Henry Parnell, *A Treatise on Roads* (London: Longman, Orme, Rees, Browne, Green and Longman, 1833); Select Committee on Acts Regarding Turnpike Roads and Highways in England and Wales, *Report*, 1819 (509).
36. Pagan, *Road Reform*, 169–172. In 1845, Pagan, an advocate of centralizing the roads under the Post Office, pointed out how frequent toll bars encumbered commerce, making it "quite impracticable to set foot upon the road at all." Pagan gave figures proving that the administration of tollgates sucked up more money than that spent on road repair. Pagan, *Road Reform*, 217.
37. "Word on the Roads," 242–243. In 1851, the Chartist William Linton voiced his call for a centralized reform of turnpikes. Arguing in the name of self-determination, individual rights, and democracy, Linton proposed the abolition of all tolls and highway rates and their replacement with a single direct tax, recommending that the state appropriate all turnpikes, rails, and canals. Linton, *Threescore and Ten Years*, 89–91.
38. Select Committee on Acts Regarding Use of Broad Wheels, *Third Report*, 1809 (271).
39. John Holt, *General View of the Agriculture of the County of Lancaster* (London: Printed for G. Nicol, 1795), 195.
40. O'Brien, *Principles of Government*, 223–224.
41. Richard Bayldon, *Sheffield and Wakefield Turnpike Road* (Barnsley: J. Ray, 1836), 4. Robert Fuge of the Bristol roads testified that "the impost of tolls is unequal and burthensome on the public, as well as on particular interests," and argued that "the relief of tolls would be highly beneficial to all classes of the community." Interview, Mr. Robert Fuge, April 21, 1836, in Select Committee on Turnpike Trusts and Tolls, *Report*, 1836 (547), 20.
42. "The facility of communication in every country is one of the causes, and also a result, of national prosperity," the committee explained. "Whatever charge or tax has a tendency to check this facility of communication ought to be avoided by the Legislature." Select Committee on Turnpike Trusts and Tolls, *Report*, 1836 (547), iii.

43. *Times*, October 25, 1856, quoted in Searle, *Turnpikes and Toll-bars*, 679. Charles Penfold explained in 1836 that local zones of commerce engendered frequent exchanges such that tolls caused a "delay and inconvenience" of particular "consequence to the parties" that "pass and repass the tollgates oftener in the year than anywhere else, and consequently are more incommoded by the interruption." Interview, Charles Penfold, May 9, 1836, in Select Committee on Turnpike Trusts and Tolls, *Report*, 1836 (547), 56. The *British Farmer's Magazine* in 1863 thought that "there is no tax which is more thoroughly offensive than the turnpike toll." It explained, "It is not the amount—though there are neighbourhoods where the amount itself becomes of consequence—it is the vexatious nature of the payment." It impugned "stopping when you are in a hurry, the fumbling in your pocket for a coin, the need of having to stop and bellow"; in short, British roads had been designed to "ruffle the temper" and impede spontaneous commerce. "The Turnpike Gate," *British Farmer's Magazine* 44 (1863): 401.
44. "Turnpike Roads and Bridges," *Hansard's Parliamentary Debates* (May 15, 1860), 1299.
45. Select Committee on Turnpike Trusts and Tolls, *Report*, 1836 (547), 6.
46. Interview, Sir James McAdam, June 20, 1836, Select Committee on Turnpike Trusts and Tolls, *Report*, 1836 (547), 123. Voices in the 1836 committee recorded that the tolls were "vexatious" to their users in setting up a minimum investment of time and money that prohibited the smallest transactions and were "expensive in the collection by the number of Collectors and Tollgates kept up." Interview, Mr. Robert Pitcher, June 6, 1836, in Select Committee on Turnpike Trusts and Tolls, *Report*, 1836 (547), 116. While I have used Macadam for the father, the son preferred to refer to himself as McAdam.
47. The numbers on parish highways had actually been gathered a year earlier by the Parliamentary Committee on Local Taxation, August 1839.
48. Bayldon, *Sheffield and Wakefield Turnpike Road*, 4.
49. Lowther reasoned that "if even one-sixth of the money collected at the turnpike gates were expended in a proper manner," London would be improved. He pointed out that the three London corridors to the north of London were governed, over a stretch of three miles, by "three acts of parliament, three sets of commissioners, and ten turnpike gates." He believed that the turnpikes were draining road-improvement fees to finance the trust's annuities and thus were enriching the trustees. Joseph Hume declared "his personal wish . . . to see every turnpike within three miles of the bridges removed." He estimated that £140,000, or two-thirds of the turnpikes' budget, was annually wasted on the managerial funds of the turnpikes themselves rather than being used to improve the roads. Mr. Maberly seconded any efforts that might save such "an

immense deal of money" for the public. A few months later, the committee demonstrated that "there can be no doubt that the present system of management pursued under these trusts is one which ought to be terminated as soon as possible." "Turnpike Trusts," *Parliamentary History and Review*, February 17, 1826, 584–585; "Turnpikes," *New Annual Register*, July 1825, 702.

50. The Metropolis Roads Trust, established by Parliament but composed voluntarily of local governments and financed out of their funds, assumed control over local turnpikes by agglomerating their inherited debts, rationalizing them, consolidating their creditors, reducing their yearly rate of interest from 8.5 percent to 5.7 percent, and setting up a sinking fund to liquidate all debts by 1850. William Albert, "The Metropolis Roads Commission: An Attempt at Turnpike Trust Reform," *Transport History* 4 (1971): 225–244; Commissioners of Metropolitan Turnpike Roads North of Thames, *First Report*, 1826–1827 (339).

51. Another metropolitan consolidation of trusts under local government was attempted in Edinburgh in the 1850s. John Sinclair, *Case for the Extension of the Municipal Boundary of Edinburgh and the Transference of the Powers of the Police and Paving Roads to the Town Council* (Edinburgh: A. and C. Black, 1855). Five Scottish counties obtained their own acts for abolishing tolls by 1870. In 1872, the Scottish National Toll Association began to clamor for toll abolition on the basis of local rule, arguing that turnpike trusts that stretched between counties necessarily took power over the roads out of the hand of the local communities best situated to run them. It cited the principle that "those should pay for the roads who used them." In 1876, Duncan McLaren led representatives for the fifteen counties to ask for toll abolition in Parliament, and the cause was granted. John Beveridge Mackie, *The Life and Work of Duncan McLaren* (Edinburgh: T. Nelson and Sons, 1888), 112, 180–190; *Hansard's Parliamentary Debates*, vol. 216 (1873), 806.

52. Thomas Penson, an authority on road administration, helped prepare a "statement of the amount of debts, the distances and the average expense of repairs which had been incurred upon these individual portions of road as well as the general expenditure of the district formerly comprised under the Montgomery Act," and on the basis of this report, a fair subdivision of debts and mortgages on the turnpike trust was prepared, and the debts of the turnpikes were divided to be paid off by more equal tolls gathered from each of the four districts of the county. Mortgagees, trustees, and local government unanimously favored this voluntary settlement, and the proposal was submitted by the county to Parliament for approval, which was effected in 1835. Interview, Thomas Penson, June 9, 1836, Select Committee on Turnpike Trusts and Tolls, *Report*, 1836 (547), 122.

53. "Parliamentary Speeches," *Times*, May 7, 1816, 2b; "Improvement of the Northern Roads," *Hansard's Parliamentary Debates* (June 3, 1830), 1337–1338.
54. F. E. D. Binney, *British Public Finance and Administration, 1774–92* (Oxford: Clarendon Press, 1958); Oliver MacDonagh, "The Nineteenth-Century Revolution in Government: A Reappraisal," *Historical Journal* 1 (1958): 52–67 Norman Chester, *The English Administrative System, 1780–1870* (Oxford: Clarendon Press, 1981); Harling, "Parliament, the State, and 'Old Corruption.'"
55. E. Hurlbut and G. Combe, *Essays on Human Rights and Their Political Guaranties* (Edinburgh: Maclachlan, Stewart, and Co., 1847); Joshua Toulmin Smith, *Local Self-Government Un-mystified* (London: E. Stanford, 1857); Joshua Toulmin Smith, *The Parish* (London: S. Sweet, 1854).
56. "Parliamentary Speeches," *Times*, May 7, 1816, 2b.
57. "Improvement of the Northern Roads," *Hansard's Parliamentary Debates* (June 3, 1830), 1337–1338.
58. Interview, J. F. Burgoyne, April 18, 1836, Select Committee on Turnpike Trusts and Tolls, *Report*, 1836 (547), 3.
59. "Plan for Expediting the Mail from London to Edinburgh, So That It Shall Arrive at One O'clock on the Second Day," *Blackwood's Edinburgh Magazine*, November 1822, 679–680.
60. "Improvement of the Northern Roads," *Hansard's Parliamentary Debates* (June 3, 1830), 1337–1338.
61. Interview, Thomas Penson, May 12, 1836, Select Committee on Turnpike Trusts and Tolls, *Report*, 1836 (547), 63, 72–73.
62. Interview, Sir James McAdam, June 20, 1836, Select Committee on Turnpike Trusts and Tolls, *Report*, 1836 (547), 126.
63. Committee on Holyhead Roads and Harbor, *First Report*, 1810 (166), 3.
64. *Gentleman's Magazine*, January 1855, 88–89.
65. For more on Macadam and the Metropolis Turnpike Trust, see R. H. Spiro, "John Loudon Macadam and the Metropolis Turnpike Trust," *Journal of Transport History* 2 (November 1956): 207–213.
66. "Improvement of the Northern Roads," *Hansard's Parliamentary Debates* (June 3, 1830), 1337–1338.
67. Select Committee on Acts Regarding Turnpike Roads and Highways, *Report*, 1811 (240), 5.
68. Select Committee on Turnpike Trusts and Tolls, *Report*, 1836 (547), 647.
69. "History of Highways," *Edinburgh Review*, April 1864, 187.
70. Toulmin Smith, *Parish*, 364.
71. William Cobbett, *Rural Rides*, 2 vols. (London: Cobbett, 1830), 1:124, original emphasis. For further context on Cobbett's critique, see Alex Benchimol,

"William Cobbett's Geography of Cultural Resistance in *Rural Rides*," *Nineteenth-Century Contexts* 26:3 (September 2004): 257–272, esp. 262.
72. Cobbett, *Rural Rides*, 1:124,
73. Milnes, *Warning Voice of a Hermit Abroad*, 115.
74. *Lounger* 88 (October 7, 1786), 171–179. The "inclosures" to which the article refers are probably squats in commons or forests, the removal of which formed part of many agrarian reform and centralized development schemes.
75. Interview, James Brook, June 6, 1836, Select Committee on Turnpike Trusts and Tolls, *Report*, 1836 (547), 118.
76. Interview, Sir James McAdam, June 20, 1836, Select Committee on Turnpike Trusts and Tolls, *Report*, 1836 (547), 124.
77. Lowther pointed to the fact that there were "more Scotch than English Members on the Committee" and cautioned his fellow members against taking their rulings as the light of reason. Lamb warned against establishing a dangerous precedent of handouts and welfare, unwarranted by previous legislation. "Upon what principle," he asked, "were the people of the West of England, who had paid for their own roads out of their own funds, to be called upon to pay for those of the North Country? Did the noble Lord fancy that the people of the North were not quite so active, industrious, and enterprising as those of the South and West, and that, therefore, they ought to be assisted by public money?" "Improvement of the Northern Roads," *Hansard's Parliamentary Debates* (June 3, 1830), 1337–1338.
78. Philp, *History of Progress in Great Britain*, 157.
79. Joshua Toulmin Smith, *Local Self-Government and Centralization* (London: J. Chapman, 1851), 206.
80. Ibid., 184.
81. Toulmin Smith, *Parish*, 363.
82. Ibid., 5. The parish was the "chief practical sphere" where English society had evolved a visible principle of the "responsibility of each to all, and of society to each of its members," tenets that Toulmin Smith defined as the "basis of the English Constitution," as well as the "only solid and permanent basis on which a free state can rest." Ibid.
83. "The Parish of Whiteacre," he explained, "maintains its own roads, constables, and other local matters." Ibid.
84. Ibid., 362.
85. W. J. Linton, "Local Government," *English Republic*, 1851, 336.
86. O'Brien, *Principles of Government*, 19, 223–224.
87. Toulmin Smith, *Parish*, 364.
88. Hurlbut and Combe, *Essays on Human Rights*, 92, 231.
89. *Lounger* 88 (October 7, 1786): 171–179.

90. Toulmin Smith, *Parish*, 363. Parish government ensured the predominance of responsibility over questions of authority. "Each man may take in turn the office of Surveyor, or of a member of a Highway Board," he explained. That self-determination sustained "the consciousness that each man . . . is part of a larger whole," the moral awareness without which democratic participation was impossible. Ibid., 362.
91. Hurlbut and Combe, *Essays on Human Rights*.
92. W. J. Linton, "Local Government," *English Republic*, 1851, 337.
93. "Bill to Amend Laws Regarding Turnpike Roads and Highways," *Parliamentary Papers* 1809 (279), "Bill to Amend Laws Regarding Turnpike Roads and Highways," *Parliamentary Papers* 1810 (102); "Bill to Amend Laws Regarding Turnpike Roads and Highways," *Parliamentary Papers* 1810 (201); "Bill to Amend Laws Regarding Turnpike Roads and Highways," *Parliamentary Papers* 1810 (255 (2)); "Bill to Amend Laws Regarding Turnpike Roads and Highways," *Parliamentary Papers* 1810 (347), "Bill to Amend Laws Regarding Turnpike Roads and Highways," *Parliamentary Papers* 1811 (48); "Bill to Amend Laws Regarding Turnpike Roads and Highways," *Parliamentary Papers* 1811 (93).
94. Thomas Grahame, *A Treatise on Internal Intercourse and Communication in Civilized States, and Particularly in Great Britain*, 8 vols. (London: Longman, Rees, Orme, Brown, Green, and Longman, 1834), 1:21, original emphasis.
95. "Highways Bill," *Hansard's Parliamentary Debates* (August 9, 1831), 1035.
96. Pagan, *Road Reform*.
97. *Hansard's Parliamentary Debates* (1839), 150–151, 44. *The Mirror of Parliament* (1840), 951–952.
98. For instance, Robert Fuge, *An Essay on the Turnpike Roads of the Kingdom* (London: T. Hurst, 1832).
99. Lord Elcho eventually succeeded to his father's title of Earl of Wemyss. See "Wemyss (Earl of), The Right Hon. Francis Wemyss Charteris Douglas," in George Washington Moon, *Men and Women of the Time: A Dictionary of Contemporaries*, 12th ed. (London: George Routledge and Sons, 1891), 940.
100. "Road Reform," *Journal of Agriculture*, July 1855–March 1857, 598, 681.
101. D. A. Smith, "Lewis, Sir George Cornewall, second baronet (1806–1863)," *Oxford Dictionary of National Biography* (Oxford: Oxford University Press, 2004). The bill proposed the compulsory association of parishes into districts with paid surveyors and the use of these districts to administer public highways, private turnpikes, and the poor laws alike. "Proceedings in Parliament," *Gentleman's Magazine*, 1850, 305.
102. George Cornewall Lewis, *On Foreign Jurisdiction and the Extradition of Criminals* (London: J. W. Parker and Son, 1859), 4.

103. For example, Alexander Dirom, *Remarks on Free Trade and on the State of the British Empire* (Edinburgh: Cadell and Co., 1827), 40. James Butler Bryan, discussing Ireland, argued that free trade and free roads together would modernize Ireland and bring about a perfection of the arts. He maintained that free roads were the perfect reflection of the "interests of the peasant." James Butler Bryan, *A Practical View of Ireland from the Period of the Union* (Dublin: W. F. Wakeman, 1831), esp. 99, 188. The cause was taken up by others as well on the same grounds, for instance, Liberal supporter of Roman Catholic emancipation and parliamentary reform, *Letters to a Friend on the Irish Reform Bill by a Liberal Supporter of Roman Catholic Emancipation and Parliamentary Reform* ([s.l.], 1831), 32. Turnpike abolition, however, was executed in Ireland only in 1857, after the famine was in full swing.
104. Evelyn Ashley, *The Life of Henry John Temple, Viscount Palmerston: 1846–1865* (London: R. Bentley and Son, 1876), 309.
105. Archibald Alison, *History of Europe from the Fall of Napoleon in MDCCCXV to the Accession of Louis Napoleon in MDCCCLII* (New York: Harper and Bros., 1860), 58.
106. "Harvest Prospects," *Draper and Clothier* 1 (1860): 53.
107. "History of Highways," 185. The highway rates were even abused to the degree that a petition against the Highway District Act was prepared and lobbied for out of expenses charged to the parish highway rate. The editorial argued that the parishes were a zone of "personal privileges" and "responsibilities which it would be beneficial to absorb." Ibid., 186.
108. *Annual Register*, 1863, 139–140.
109. The committee also noted the inability of most parishes to employ a professional surveyor and the failure of permissive legislation to persuade many parishes to unite into districts where employing professional engineers would be simplified. It suggested the enforced "common management" of new highway districts and recommended that the "Parliamentary Commissioners should form Districts and Subdivisions throughout England and Wales." These highway districts, it hoped, would allow the more efficient extermination of the turnpike trusts, which would be consolidated and then governed according to the oversight of the local districts. Commission for Inquiry into the State of Roads in England and Wales, *Report*, 1840 (256), 12.
110. Commissioners for Inquiring into Large Towns and Populous Districts, *First Report*, 1844 (572). The same claims were made also by various reports of the Sanitary Commissioners, whose administrative districts also depended on the existence of voluntary highway districts under the 1835 act.
111. William Henry Barrow, MP for South Nottinghamshire, who led the attacks on centralized highways in the 1850s, was notorious for being among the minority who "censured free trade." *Dod's Parliamentary Companion*, 1854: 135.

112. For instance, Fuge, *Essay on the Turnpike Roads of the Kingdom.*
113. *Turnpike Trusts Consolidation Bill, Objections Thereto,* 16 May 1836 (printed), National Archives Q26/3/408 1836. My research suggests that the actual date should be 1834 instead.
114. Commission for Inquiry into the State of Roads in England and Wales, *Report,* 1840 (256).
115. "Highways," *Hansard's Parliamentary Debates* (February 13, 1850): 748–755.
116. R. H. R., "Henley, Joseph Warner," *The Oxford Dictionary of National Biography* (Oxford: Oxford University Press, 1908), 416.
117. In 1852, Mr. Barrow spoke against the bill "at some length," opposing it as "an unnecessary and unwarrantable interference with the right of property" and as an "extension of that principle of centralization and bureaucracy to which the people of this country entertained such a deep-rooted antipathy." "Highways Bill," *Parliamentary Remembrancer,* July 7, 1860, 178.
118. Although the 1840 committee was reluctant to provide a nationwide system, it voiced the imperative for "some system of consolidation" and gestured toward the countywide systems in place in Scotland and the metropolitan trusts. It noted the financial complications of dealing with both lucrative and bankrupt trusts at the same moment and recommended "the appointment of a Parliamentary Commission empowered to consider and decide on each case, reserving to each creditor the right to refer the value of his claim to a jury or to arbitration." Commission for Inquiry into the State of Roads in England and Wales, *Report,* 1840 (256), 11.
119. Milnes, *Warning Voice of a Hermit Abroad,* 22.
120. Toulmin Smith, *Parish,* 362–363.
121. The 1836 committee recorded instances of "jobbing" within the grand jury system in Ireland that governed turnpikes there. Interview, Mr. James O'Sullivan, May 2, 1836, Select Committee on Turnpike Trusts and Tolls, *Report,* 1836 (547), 47.
122. Combination occurred where trusts bidding to take over a metropolitan area might bid lower rates than the ones they ultimately planned to institute. Interview, Charles Penfold, May 9, 1836, Select Committee on Turnpike Trusts and Tolls, *Report,* 1836 (547), 57–58.
123. "Highways Bill," *Parliamentary Remembrancer,* December 12, 1857, 2. The same journal later announced "the withdrawal" of a new bill as "a useful lesson to every one" concerned with "the common interest of his country." "Highways Bill—Withdrawn," *Parliamentary Remembrancer,* July 28, 1860, 207.
124. "Highways Bill," *Parliamentary Remembrancer,* July 28, 1860, 207.
125. "Highways Bill," *Parliamentary Remembrancer,* July 7, 1860, 178. The *Edinburgh Review* characterized the centralizers as the "carriage interest," or the interests of the few represented against "the public." It summed up the contest

by explaining that the measure was opposed because "it was a centralizing measure, and calculated to increase expenditure."
126. "History of Highways," 365.
127. *Hansard's Parliamentary Debates* (1839): 150–151; *The Mirror of Parliament* (1840), 951–952.
128. "Highways," *Hansard's Parliamentary Debates* (February 13, 1850): 748–755.
129. "Highways Bill," *Parliamentary Remembrancer,* December 12, 1857, 2.
130. "Extract from Lord Brougham's Letter to the Earl of Radnor," *Law Magazine and Law Review,* 1862, 62.

4. Wayfaring Strangers

1. Trial of John Malcolm, *The Proceedings of the Old Bailey,* February 16, 1791, ref. t17910216-3. All Old Bailey references are online at http://www.oldbailey online.org/.
2. J. A. Houlding, *Fit for Service: The Training of the British Army, 1715–1795* (Oxford: Oxford University Press, 1981), 11; Richard Middleton, "The Recruitment of the British Army, 1755–1762," *Journal of the Society for Army Historical Research* 67 (1989): 226–238.
3. Clive Emsley, *British Society and the French Wars, 1793–1815* (Totowa, N.J.: Rowman and Littlefield, 1979), 133; John Brewer, *The Sinews of Power: Money, War, and the English State, 1688–1783* (Boston: Unwin, 1989), 30; Huw V. Bowen, *War and British Society, 1688–1815* (Cambridge: Cambridge University Press, 1998), 12.
4. Samuel Hutton, "Autobiography," reprinted in Roy Palmer, ed., *The Rambling Soldier: Life in the Lower Ranks, 1750–1900* (Harmondsworth: Penguin Books, 1977), 15–17.
5. Benjamin Harris, *A Dorset Rifleman: The Recollections of Benjamin Harris* (Swanage: Shinglepicker, 1996), 124.
6. Rotation was not instituted until later in the eighteenth century, so until then enrolling in the army meant being shipped overseas without relief for possibly the whole of the engagement or longer, if one stayed into peacetime. "One quarter of the marching Foot of the British Army was left to rot, unrelieved, on distant foreign stations for the whole of the period," observes Houlding, *Fit for Service,* 19.
7. Ibid., 25.
8. The main billeting towns in England were Chester, Malpas, Whitchurch, Oswestry, Shrewsbury, Church Stretton, Ludlow, Leominster, Hereford, and Ross. Regiments were almost never stationed in Wales, although small detachments were occasionally found at Carmarthen, Aberystwyth, and Aberdovey. The only exceptions were in far northern Wales, Devon, Somerset, the Pennines and Welsh Cambrians, and the Grampians and the Cheviot Hills, where population was sparse. Ibid., 11, 25, 28.

9. *The Portsmouth Guide* (Portsmouth: R. Carr, 1775), 18.
10. John Bullar, *A Companion in a Tour round Southampton* (Southampton: T. Baker, 1799), 107, 204.
11. Henry Slight and Julian Slight, *Chronicles of Portsmouth* (London: Relfe, 1828), 48.
12. Charles Dickens, *The Personal History and Experience of David Copperfield the Younger* (New York: Macmillan, 1911), 33–34.
13. Middleton, "Recruitment of the British Army," 229; Houlding, *Fit for Service*, 10, 13, 117.
14. Stephen Brumwell, "Rank and File: A Profile of One of Wolfe's Regiments," *Journal of the Society for Army Historical Research* 79:317 (2001): 3–24; Houlding, *Fit for Service*, 17.
15. Linda Colley, *Britons: Forging the Nation* (London: Pimlico, 1994); Andrew MacKillop, "'Plough-shares into Broadswords': The Scottish Highlands and War, 1750–1810," in Ewen A. Cameron and F. Watson, eds., *Scotland and War: Annual Conference, Association of Scottish Historical Studies* (Perth: Association of Scottish Historical Studies, 1995), 52–61; Andrew MacKillop, *"More Fruitful than the Soil": Army, Empire and the Scottish Highlands, 1715–1815* (East Linton: Tuckwell, 2000); Steve Murdoch and Andrew MacKillop, *Fighting for Identity: Scottish Military Experience, c. 1550–1900* (Boston: Brill, 2002). With the Militia Act of the 1790s, quotas for recruitment to the militia were extended to Scotland for the first time. Middleton, "Recruitment of the British Army," 231; E. M. Lloyd, "The Raising of the Highland Regiments in 1757," *English Historical Review* 17 (1902): 466–469; Benjamin Harris, *Dorset Rifleman*, 19; J. E. Handley, *The Irish in Scotland* (Cork: Cork University Press, 1945), 165; Eric Richards, *A History of the Highland Clearances* (London: Croom Helm, 1982); Bowen, *War and British Society*.
16. John Wiliam Fletcher was a military officer from a military family in Switzerland before he met the Wesleys in the 1750s; he later carried Methodist teachings to Wales and Switzerland. Captain Gallantine, a Methodist army officer stationed at Musselburgh, was the correspondent who initially persuaded the Wesleys to visit Scotland in 1751. Henry D. Rack, *Reasonable Enthusiast: John Wesley and the Rise of Methodism* (Philadelphia: Trinity Press International, 1989), 229.
17. Thomas Percival Bunting, *The Life of Jabez Bunting: With Notices of Contemporary Persons and Events* (New York: Harper and Brothers, 1859), 71.
18. D. Caroline Hopwood, *An Account of the Life and Religious Experiences, of D. Caroline Hopwood* (Leeds: Printed by E. Baines, 1801).
19. Robert Philip, *The Life and Times of the Reverend George Whitefield, M.A.* (New York: D. Appleton and Co., 1838), 65.
20. Ibid., 308.

21. Matthew Richey, *A Memoir of the Late Rev. William Black, Wesleyan Preacher* (Halifax, N.S.: William Cunnabell, 1839); Nathan Bangs, *The Life of the Rev. Freeborn Garrettson* (New York: J. Emory and R. Waugh, 1820).
22. Philip, *Life and Times of the Reverend George Whitefield*, 68. See also Robert Adam, *The Religious World Displayed; or, A View of Judaism, Paganism, Christianity and Mohammedanism* (London: Longman, Hurst, Rees, and Orme, 1809), 98–101; Joshua Marsden, *The Narrative of a Mission to Nova Scotia, New Brunswick, and the Somers Islands* (Plymouth Dock: Printed and sold by J. Johns, 1816), 254–255.
23. *London Christian Instructor or Congregational Magazine* 7 (1824): 673.
24. H. L. Hughes, *Life of Howell Harris* (London: J. Nisbet and Co., 1892), 279; Joseph Evans, *Biographical Dictionary of Ministers and Preachers of the Welsh Calvinistic Methodistic Body* (Carnarbon: D. O'Brien Owen, 1907), 137.
25. "King v. Churchyard and Others," *Edinburgh Annual Register* 4:2 (1811): 36–39.
26. Similar accounts have been accumulated by W. H. Fitchett, "Soldier Methodists," chap. 9 in Fitchett, *Wesley and His Century, a Study in Spiritual Forces* (Nashville: Smith, Lamar, 1906), 225–239; and John Telford, *Wesley's Veterans: Lives of Early Methodist Preachers Told by Themselves* (London: Robert Culley, 1912).
27. John Wesley, *The Works of the Rev. John Wesley* (London: Sold by Thomas Blanshard, 1809), 2:413, 5:518.
28. "On the Death of Mr. Whitefield," in Wesley, *Works*, 2:4.
29. Howell Harris, *Visits to London* (Aberystwyth: Cambrian New Press, 1960), 4–5.
30. John Greaves Nall, *Great Yarmouth and Lowestoft* (London: Longman, Green, Reader, and Dyer, 1866), 113.
31. Howell Harris, *A Brief Account of the Life of Howell Harris, Esq* (Trevecka, 1791), 93; John Chambers, ed., *A General History of the County of Norfolk* (Norwich: J. Stacey, 1829), 1336.
32. John Wesley, *An Extract of the Rev. Mr. John Wesley's Journal* (London: New Chapel, 1788), 47.
33. John Pawson, *The Letters of John Pawson: Methodist Itinerant, 1762–1806*, 3 vols. (Peterborough: Published on behalf of the World Methodist Historical Society, 1995), 1:36.
34. David Hempton, *The Religion of the People: Methodism and Popular Religion, c. 1750–1900* (New York: Routledge, 1996), 131.
35. "John Haime," in John Wesley and John Pawson, *The Experience of Several Eminent Methodist Preachers* (Joseph Dix, 1812), 36.
36. John Haime, *A Short Account of God's Dealings with Mr. John Haime* (London: J. Paramore, 1785); "John Haime," in John Wesley and John Pawson, *The Experience of Several Eminent Methodist Preachers* (Chambersburg, Pa.:

Printed for Thomas Yeats and Thomas Johns, by G. K. Harper, 1812), 25–52; Robert Southey, *The Life of Wesley; and the Rise and Progress of Methodism*, 2 vols. (York: Evert Duyckinck and George Long, 1820), 2:57–63.

37. Abel Stevens, *The History of the Religious Movement of the Eighteenth Century, Called Methodism* (New York: Carlton and Porter, 1859), 330–331; Richard Douglas Moore, *Methodism in the Channel Islands* (London: Epworth Press, 1952), 18.
38. "Methodism in the Channel Islands," *Wesleyan-Methodist Magazine* 16 (1870): 62; Moore, *Methodism in the Channel Islands*, 17.
39. "Obituary—William Strutt, Esq., F.R.S.," *Monthly Repository of Theology and General Literature*, 50, February 1831, 138; Rack, *Reasonable Enthusiast*, 279.
40. Howell Harris, *Brief Account*, 54–57.
41. Robert Southey, in his 1820 *Life of Wesley*, dedicated a chapter in his section on the history of Christianity to the traditional confinement of the faith to cities. In this narrative, Wesley's revolutionary role for the religion was to stretch the geographic bounds of Christianity. Southey, *Life of Wesley*, 1:190.
42. Francis Fletcher Bretherton, *Early Methodism in and around Chester, 1749–1812* (Chester: Phillipson and Golder, 1903), 3.
43. Jonathan Crowther, *The Life of Thomas Coke* (Leeds: Alexander Cumming, 1815), 537.
44. "Review of the Life of Dr. Adam Clarke," *Christian Observer and Advocate*, November 1833, 686.
45. Southey, *Life of Wesley*, 2:28–29.
46. John Taylor, *The Apostles of Fylde Methodism* (London: T. Woolmer, 1885), 33.
47. Thomas Shaw, *History of Cornish Methodism* (Truro: Barton, 1967), 51.
48. Cornelius Cayley, *The Riches of God's Free Grace, Display'd in the Conversion of Cornelius Cayley* (Norwich, 1757), 109.
49. Rack, *Reasonable Enthusiast*, 243, 245, 212, 217, 229; Stevens, *History of the Religious Movement*, 190; Richard Watson, *The Life of Rev. John Wesley* (London: John Mason, 1831), chap. 10; John B. Dyson, *The History of Wesleyan Methodism in the Congleton Circuit* (London: H. Mason, 1856), chap. 3; Benjamin Ingham, *Diary of an Oxford Methodist* (Durham, N.C.: Duke University Press, 1984); Charles Henry Crookshank, *History of Methodism in Ireland* (London: T. Woolmer, 1886).
50. No societies existed along the channel coast or in East Anglia, the eastern Midlands, most of Wales, Yorkshire, and Ireland. Scotland had no preachers at all.
51. R. Heitzenrater, *The Elusive Mr. Wesley* (Nashville: Abingdon Press, 1984), 162, 181, 190, 217; Rack, *Reasonable Enthusiast*, 229.
52. C. Yrigoyen and S. E. Warrick, *Historical Dictionary Of Methodism* (London: Scarecrow Press, 2005), 169.

53. Moore, *Methodism in the Channel Islands*, 35–36.
54. Bunting, *Life of Jabez Bunting*, 80–81.
55. Shaw, *History of Cornish Methodism*, 53.
56. Bretherton, *Early Methodism in and around Chester*, 26.
57. Taylor, *Apostles of Fylde Methodism*, 14.
58. Dawson Turner, *Sepulchral Reminiscences of a Market Town* (Yarmouth: Charles Barber, 1848), 151.
59. Nathan Bangs, *The Life of the Rev. Freeborn Garrettson* (New York: J. Emory and R. Waugh, 1820), 166.
60. *Methodist Magazine* 2 (1799): 511.
61. Rupert Eric Davies, *A History of the Methodist Church in Great Britain*, 4 vols. (London: Epworth, 1965), 1:236.
62. Shaw, *History of Cornish Methodism*, 23. Contemporaries give rather more exaggerated figures, which may or may not reflect occasional attendants. Crowther argues that by 1814 there were 173,885 Methodists in Great Britain (excluding Ireland), served by 685 regular traveling preachers, and nearly equal numbers throughout the greater worldwide church. Crowther, *Life of Thomas Coke*, 486.
63. T. W. Laqueur, *Religion and Respectability: Sunday Schools and Working Class Culture, 1780–1850* (New Haven, Conn.: Yale University Press, 1976).
64. John Cennick, *Discourses on Important Subjects* (London: M. Lewis, 1762), xxiii; Frank Baker, *John Cennick, 1718–55: Handlist of His Writings* (Leicester: A. A. Taberer, 1958).
65. Southey, *Life of Wesley*, 2:48.
66. Ibid., 2:52.
67. Bretherton, *Early Methodism in and around Chester*, 26.
68. Rack, *Reasonable Enthusiast*, 218; "Memoir of the Rev. John Bowers," *Wesleyan-Methodist Magazine* 16 (1870): 194.
69. Ingham, *Diary of an Oxford Methodist*; Elie Halévy, *The Birth of Methodism in England* (Chicago: University of Chicago Press, 1971), 71; Stevens, *History of the Religious Movement*, 337.
70. Rack, *Reasonable Enthusiast*, 221; Shaw, *History of Cornish Methodism*, 14.
71. James Sigston, *Memoir of the Life and Ministry of Mr. William Bramwell* (London: J. Nichols, 1820); Taylor, *Apostles of Fylde Methodism*, 30–37.
72. Jacob Halls Drew, *The Life, Character, and Literary Labours of Samuel Drew, A. M.* (London: Longman, Rees, Orme, 1834); Shaw, *History of Cornish Methodism*, 56.
73. Turner, *Sepulchral Reminiscences of a Market Town*, 151.
74. John Wesley, *An Extract of the Reverend Mr. John Wesley's Journal, from July XX, 1750* (London: Printed and sold at the Foundery, 1759), 46.
75. Adam, *Religious World Displayed*, 100.

76. Nathan Bangs, *A History of the Methodist Episcopal Church* (New York: T. Mason and G. Lane, 1840), 255–260.
77. The Methodist leadership responded by reemphasizing conversion-centered missionaries, oriented to pushing the network of circuits deep into yet-unvisited regions, both at home and abroad. The interdenominational London Missionary Society was founded in 1795. In 1805, the first Wesleyan home mission stations were set up. Pawson, *Letters of John Pawson*, 2:129, n. 259.
78. Ibid., 2:144. Even if itinerancy was on the wane, Methodists had established a culture of geographic penetration that would survive. By the 1790s, John Pawson was often reflecting on the competition with evangelicals, and thus the necessity of Methodists extending their geographic reach in order to remain a force in the religion market. In a letter to Joseph Benson, pastor of the Methodist Chapel at Hull, Pawson wrote: "I was a good deal struck in reading the Evangelical Magazine the other day to see how different bodies of Dissenters are uniting in order to send Missionaries almost over all the world, and also in the darkest parts of this nation. All these have lighted their candle at the Methodist lamp. But if we do not take great care, they will rob us of our glory." Ibid., 2:144. Pawson's concern reflected the challenges posed by rival evangelical churches that had begun to copy Wesley's pattern of itinerancy. The Village Itinerant Society was founded in 1796 by John Eyre, editor of the *Evangelical Magazine*.
79. John Petty, *The History of the Primitive Methodist Connexion* (London: R. Davies, 1860), 572–574.
80. John Rule, *The Labouring Classes in Early Industrial England, 1750–1850* (London: Longman, 1986).
81. A survey of one of Wolfe's regiments shows that 79 percent of recruits from Ireland in 1757 were younger than thirty. Stephen Brumwell's survey of one of Wolfe's regiments in the Seven Years' War showed that harvest laborers were the largest group by profession, 35 percent; weavers were the next largest, 17 percent; after them came shoemakers, 7 percent, and tailors, 5 percent. Brumwell, "Rank and File," 3–24. The average age of enlistment among the troops at Louisbourg was 22.6 years. Houlding, *Fit for Service*, 13.
82. Hutton, "Autobiography," 15–17.
83. John Brown, *Sixty Years' Gleanings from Life's Harvest: A Genuine Autobiography* (Cambridge: J. Palmer, 1858), 43–45.
84. Benjamin Harris, *Dorset Rifleman*, 70, 136.
85. Humphrey Southall, "Mobility, the Artisan Community and Popular Politics in Early Nineteenth-Century England," in Gerry Kearns and Charles W. J. Withers, eds., *Urbanising Britain* (Cambridge: Cambridge University Press, 1991), 119.

86. W. H. Oliver, "The Labour Exchange Phase of the Co-operative Movement," *Oxford Economic Papers* 10:3 (1958): 355–367; Robert Iliffe, "Material Doubts: Hooke, Artisan Culture and the Exchange of Information in 1670s London," *British Journal for the History of Science* 28 (1995): 285–318; E. J. Hobsbawm, "The Tramping Artisan," *Economic History Review* n.s., 3:3 (1951): 299–320; Humphrey Southall, "Towards a Geography of Unionization: The Spatial Organization and Distribution of Early British Trade Unions," *Transactions of the Institute of British Geographers* 13:4 (1988): 466–483.

87. By 1741, weavers had a tramping system in place that derived from a more local one set up in 1708. Curriers had one by 1750 and hatters by the 1770s. By the end of the century, papermakers and calico printers were well organized. Increasingly, friendly societies of tradesmen were organized to save traveling journeymen from bottoming out on the tramp. From their onset, dues of local friendly societies had funded revelrous fellowship and had protected the sick and aged; by the 1790s, most friendly societies also collected dues for tramping artisans. J. Rule, *The Experience of Labour in Eighteenth-Century Industry* (London: Croom Helm, 1981), 152ff.

88. "Indeed, the village cobbler was often the village atheist." Reginald J. White gives the example of Thomas Hardy Sr., William Benbow of the General Strike, and one of the ten Peterloo prisoners. Reginald J. White, *Waterloo to Peterloo* (London: William Heinemann, 1957), 20.

89. H. J. H. Gosden, *The Friendly Societies in England, 1815–75* (Manchester: Manchester University Press, 1961), chap. 1; London Metropolitan Archive 4254.

90. Brown writes, "Each member pays a certain amount monthly, in order to raise a fund for the support of families when a strike takes place, whether in one shop or more as the case may be: the men being at the same time furnished with 'tramping money,' to enable them to go into the country until the dispute betwixt employer and employed is for the time adjusted. All persons subscribing to this fund are called 'flints,' whilst those who do not join are honoured with the dignified title of 'scabs.'" Tramping money was twelve to fourteen shillings from the general fund to start them on a journey. Brown, *Sixty Years' Gleanings from Life's Harvest*, 42–43.

91. In his survey of the travels of steam engine makers in the 1840s, Southall finds remarkable rates of mobility among a significant proportion of skilled laborers. An average of 13 percent of the membership went on the tramp each year during the period after 1835, with an increase during the 1842 depression. As Southall explains, the tramping system turned the youngest and most vulnerable members of trade society out on the road. If they were lucky, they could set up and make their living for a while in the next town or farther afield. If they were unlucky, or if the dearth was more general, no job might be had

anywhere, in which case the tramping journeymen fell down through the ranks of the protected trades to the general pool of mobile poor, reliant on the informal labor sector, poor relief, charity, or military recruitment to save them from starvation. Southall, "Towards a Geography of Unionization," 466–483.
92. Rule, *Experience of Labour in Eighteenth-Century Industry*, 152ff.
93. The Victory Lodge in Manchester was founded in 1809 by Bolton, a marble cutter who had traveled from London to Manchester. The Salford Lodge was founded in 1810, the Manchester Unity around 1814, and the Loyal Abercrombie Lodge of London in 1822. J. Burn, *An Historical Sketch of the Independent Order of Oddfellows* (Manchester: A. Heywood, 1845), 26–27.
94. By the 1840s, many trades, such as the stonemasons, offered not only tramping money but a reserved bed in a public house, manned by a relieving officer and bed inspector. Southall, "Mobility," 105, 120.
95. The highest rates of membership were among workers in woolen/cotton manufactures, miners, shoemakers, carpenters/joiners, and workers in the building trades; shoemaking, carpentry, and building were among the most seasonally fluctuating and therefore mobile industries. Gosden, *Friendly Societies in England*, chap. 3.
96. Burn, *Historical Sketch of the Independent Order of Oddfellows*, 35.
97. Notable Chartists, including Thomas Cooper and John Skevington, Leicestershire Chartists in 1839–1848, were former Methodist preachers; Skevington was a former itinerant among the Primitives. Methodist preachers supported the charter and opened their chapels to Chartist meetings throughout the north. From 1839 forward, Chartists led summer camp meetings in direct imitation of the mobile Methodist camps. R. F. Wearmouth, *Methodism and the Working-Class Movements of England, 1800–1850* (London: Epworth Press, 1937); Wearmouth, *Methodism and the Common People of the Eighteenth Century* (London: Epworth Press, 1945); Wearmouth, *Methodism and the Trade Unions* (London: Epworth Press, 1959).
98. Peter Clark, "Migrants in the City," in Peter Clark and David Souden, eds., *Migration and Society in Early Modern England* (Totowa, N.J.: Barnes and Noble, 1987), 267–291; I. D. Whyte, *Migration and Society in Britain, 1550–1830* (Basingstoke: Macmillan, 2000).
99. Arthur Redford, *Labour Migration in England, 1800–1850* (Manchester: Manchester University Press, 1964), especially chap. 4.
100. Colin Pooley and Jean Turnbull, *Migration and Mobility in Britain since the 18th Century* (London: University College London Press, 1998).
101. Previous generations believed in a massive upward shift in mobility during the eighteenth century as the result of agrarian revolution and demographic change, but recent studies have drastically revised those findings. We now identify how the same short-distance movements gradually pushed individuals

toward cities without observing a major shift in the distances traveled by any single generation. Redford, *Labour Migration in England*; Jan Lucassen, *Migrant Labor in Northern Europe, 1600–1900* (London: Croom Helm, 1987), 29–30, 32, 34–35, 40; Malcolm Kitch, "Population Movement and Migration in Pre-industrial Rural England," chap. 4 in Brian Short, ed., *The English Rural Community* (Cambridge: Cambridge University Press, 1992), 62–84.

102. James Roderick O'Flanagan, *Impressions at Home and Abroad* (London: Smith, Elder, and Co., 1837), 84.

103. "The Night Walker," *Blackwood's Edinburgh Magazine*, 1823, 509; Charles Dickens, *Oliver Twist* (Philadelphia: Lea and Blanchard, 1839), 86; George Mogridge, *Old Humphrey's Walks in London and Its Neighbourhood* (London: Religious Tract Society, 1839), 345.

104. Dickens, *Oliver Twist*, 86. Detailed figures on the amount of food daily carried to the metropolis on human backs may be found in "Review of Potts' Gazetteer of England and Wales," *Monthly Review*, 1811, 310.

105. Audrey Eccles, "Vagrancy in Later Eighteenth-Century Westmorland: A Social Profile," *Transactions of the Cumberland and Westmorland Antiquarian and Archaeological Society* 89 (1989): 252; M. Dorothy George, *London Life in the XVIIIth Century* (London: Kegan Paul, Trench, Trubner, 1925), 124.

106. Peter Leese, *The British Migrant Experience, 1700–2000: An Anthology* (New York: Palgrave, 2002), 45–46; Lionel Rose, *Rogues and Vagabonds: Vagrant Underworld in Britain, 1815–1985* (London: Routledge, 1988), chap. 1.

107. Pooley and Turnbull, *Migration and Mobility in Britain*; D. Friedlander and R. J. Roshier, "A Study of Internal Migration in England and Wales: Part I," *Population Studies* 19:3 (1966): 239–379; R. Lawton, "Population Changes in England and Wales in the Later Nineteenth Century: An Analysis of Trends by Registration Districts," *Transactions of the Institute of British Geographers* 44 (1968): 55–74; M. J. Greenwood and L. B. Thomas, "Geographic Labor Mobility in Nineteenth Century England and Wales," *Annals of Regional Science* 7:2 (1973): 90–105; Peter Clark, "Migration in England during the Late Seventeenth and Early Eighteenth Centuries," *Past and Present* 83 (1979): 59; R. Lawton, "Mobility in Nineteenth-Century British Cities," *Geographical Journal* 145:2 (1979): 206–224; Stephen Nicholas and Peter R. Shergold, *Internal Migration in England, 1817–1839* (Canberra: Australian National Library, 1985); S. Nicholas and P. R. Shergold, "Internal Migration in England, 1818–1839," *Journal of Historical Geography* (1987): 163–164; Martin B. White, "Family Migration in Victorian Britain: The Case of Grantham and Scunthorpe," *Local Population Studies* 41 (1988): 41–50; Whyte, *Migration and Society in Britain*; J. Long, "Rural-Urban Migration and Socioeconomic Mobility in Victorian Britain," *Journal of Economic History* 65:1 (2005): 1–35.

108. Rule, *Labouring Classes in Early Industrial England*.

109. Eccles, "Vagrancy in Later Eighteenth-Century Westmorland," 251.
110. Ibid., 261.
111. Leese, *British Migrant Experience*, 40–41.
112. [Mary Saxby], *Memoirs of a Female Vagrant* (London: J. Burditt, 1806).
113. Jem Mace, *Fifty Years a Fighter* (Chippenham, Wiltshire: Caestus Books, 1998).
114. In Eccles's study of vagrants in Westmorland, the numbers of tinkers and peddlers convicted as criminals surpassed those of all other groups except veterans and textile workers. Eccles, "Vagrancy in Later Eighteenth-Century Westmorland," 261.
115. Leese, *British Migrant Experience*, 43–44.
116. [Saxby], *Memoirs of a Female Vagrant*.
117. Stephen Allott, *Friends in York: The Quaker Story in the Life of a Meeting* (York: Sessions, 1978), 31. Her story is also told in part in John Conran, *A Journal of the Life and Gospel Labours of John Conran* (London: Charles Gilpin, 1850), 38–47.
118. Hertfordshire Record Office, D/Ese F20.
119. Long-distance travel by women preachers was replaced by shorter-distance preaching by informal women preachers, who tended to persist in communities of Primitive Methodism where the established church had less of a presence. "Memoir of Elizabeth Tripp," *Primitive Methodist Magazine* 8 (1838): 216.
120. Eccles, "Vagrancy in Later Eighteenth-Century Westmorland," 251.
121. "Trial of James Barrett," *The Proceedings of the Old Bailey*, July 7, 1779, ref. t17790707-49.
122. "Trial of Robert Parker," *The Proceedings of the Old Bailey*, October 24, 1770, ref. t17701024-22; "Trial of Samuel Bacon," *The Proceedings of the Old Bailey*, October 16, 1751, ref. t17511016-3; "Trial of James Rockett, otherwise Price, Timothy Steward," *The Proceedings of the Old Bailey*, February 22, 1764, ref. t17640222-66.
123. David A. Kent, "'Gone for a Soldier': Family Breakdown and the Demography of Desertion in a London Parish, 1750–91," *Local Population Studies* 45 (1990): 27–42; Peter Drake, *The Memoirs of Capt. Peter Drake* (Dublin: S. Powell, 1755), 103–104.
124. Southall, "Mobility," 116; Iorwerth J. Prothero, *Radical Artisans in England and France, 1830–1870* (Cambridge: Cambridge University Press, 1997), 55.
125. C. M. Clode, *The Military Forces of the Crown: Their Administration and Government*, 2 vols. (London: J. Murray, 1869), 1:229–238.
126. Palmer, *Rambling Soldier*, 19.
127. Drake, *Memoirs of Capt. Peter Drake*, 138–139.
128. Thomas Simes, *The Military Medley* (Dublin: S. Powell, 1767), 234.
129. Stephen Brumwell, *Redcoats: The British Soldier and War in the Americas, 1755–1763* (Cambridge: Cambridge University Press, 2006).

130. The Royal Navy was reduced from 7,600 to 1,750 within a year after the end of hostilities in 1763. Michael Duffy, "The Foundations of British Naval Power," in Duffy, ed., *The Military Revolution and the State, 1500–1800* (Exeter, 1980), 49–85, esp. 51–53.
131. In Stephen Brumwell's study of the Seven Years' War, he found demobilization rates of 26 percent within one month. In one regiment with 117 discharges, only 31 were recommended for Chelsea pensions; the rest were on their own dole. Brumwell, "Rank and File," 3–24.
132. Houlding, *Fit for Service*, 13–14.
133. Audrey Eccles's survey of vagrants in eighteenth-century Westmorland found that the highest occupation by far of those prosecuted was military: 80 out of 202 cases involved veterans of the army or navy. Eccles, "Vagrancy in Later Eighteenth-Century Westmorland," 261.
134. Brumwell, *Redcoats*.
135. J. M. Beattie, "The Pattern of Crime in England, 1660–1800," *Past and Present* 62 (1974): 47–95; P. Lawson, "Property Crime and Hard Times in England, 1559–1624," *Law and History Review* 4:1 (1986): 95–127.
136. "Trial of Richard Dennison," *The Proceedings of the Old Bailey*, May 15, 1755, ref. t17550515-20
137. The pattern of eighteenth-century trials examined by J. M. Beattie demonstrates that violent crimes escalated at the end of every war, with discharged veterans outnumbering all other classes of criminals. Beattie, "Pattern of Crime in England."
138. *Notices of the Labours of the Rev. John Cennick*, ix.
139. Heitzenrater, *Elusive Mr. Wesley*, 132–133. John and Charles Wesley recorded in their diaries a "war against the Methodist" waged throughout the 1750s in the form of yelling, bell ringing, and organ playing during their preaching, as well as attacks with stones, eggs, and tomatoes. John Cennick was attacked by mobs in Swindon, Stratton, and Lyneham. John Wesley had a reputation as an Oxonian that brought him grief in his journeys through Cornwall; Charles Wesley was charged in Yorkshire with treasonable speech on the part of the Jacobite rebels. Ibid., 151.
140. Apparently the justices of the peace had offered Maxfield to a man-of-war, the captain of which rebuked them by saying, "I have no authority to take such men as these, unless you would have me give him so much a week to preach and pray to my people." John Wesley, *The Journal of the Reverend John Wesley*, 2 vols. (New York: Carlton and Phillips, 1855), 1:338.
141. John Gillies and George Whitefield, *Memoirs of Rev. George Whitefield* (Middletown: Hunt and Noyes, 1838), 100; Erasmus Middleton, *Evangelical Biography: Being a Complete and Fruitful Account of the Lives* (London: Printed for J. Stratford, 1807), 225.

142. Southey, *Life of Wesley*, 2:50.
143. Moore, *Methodism in the Channel Islands*, 21.
144. Robert R. Wark, *Rowlandson's Drawings for a Tour in a Post Chaise* (San Marino, Calif.: Huntington Library, 1963).
145. In the 1730s, "flying wagons" began to compete with "flying coaches" for the carrying of heavy goods. The first were initiated by Walter Wiltshire, whose "Wiltshire's Flying Waggons" established a weekly pattern of journeys between Bath and London. W. T. Jackman, *The Development of Transportation in Modern England* (Cambridge: University Press, 1916); J. A. Chartres, "Road Carrying in England in the Seventeenth Century: Myth and Reality," *Economic History Review* 30:1 (1977): 73–94; G. Adams, "The Coach Motif in Eighteenth-Century Fiction," *Modern Language Studies* 8:2 (1978): 17–26; C. H. Wilson, "Land Carriage in the Seventeenth Century," *Economic History Review* 33:1 (1980): 92–95.
146. As wagons carried goods to new parts of the kingdom, so too did stagecoaches use the new road network to carry people across the counties. At the beginning of the century, a slow journey was established three times a week between London and several major hubs, such as York and Exeter. More reliable road surfaces lent themselves to more efficient travel, and between 1730 and 1750 faster "flying" coaches began traveling more regular, direct circuits. The new coach routes reached deeper into the northern periphery. In 1731, Nicholas Rothwell of Warwick set up the first stagecoach between Birmingham and London. Howard Robinson, *The British Post Office* (Princeton, N.J.: Princeton University Press, 1948), 128; Stella Margetson, *Journey by Stages: Some Account of the People Who Travelled by Stage-coach and Mail in the Years between 1660 and 1840* (London: Cassell, 1967). The first Glasgow and Edinburgh connection by stage appeared in 1749. In 1754, the first Edinburgh-London stagecoach was opened by Hosea Eastgate of Soho. Better vehicles held up against the wear of distance and velocity. Within ten years of the route's introduction, between 1749 and 1759, the "Glasgow and Edinburgh Fly" halved the speed of the "Glasgow and Edinburgh Caravan" from four days to two. Similar improvements proliferated elsewhere after 1750. The ride from London to York took four days in 1700 and 1750, but only three by 1784. Robinson, *British Post Office*, 129–130; Jackman, *Development of Transportation in Modern England*, 96, 137, 285–286; Sidney Webb and Beatrice Webb, *The Story of the King's Highway* (London: Longmans, Green and Co., 1913), 72, chaps. 5 and 7.
147. The first established coach route left Bristol at 4 P.M. and arrived in London at 8 A.M., taking one hour less on the road than the stagecoach route. On longer routes, the time difference was greater. The journey to Scotland was more than halved, from 135 hours to 60. Robinson, *British Post Office*, 139; C. Noall,

A *History of Cornish Mail- and Stage-coaches* (Truro: D. Bradford Barton, 1963); B. Austen, *British Mail-coach Services, 1784–1850* (doctoral diss., University of London, 1979).

148. Iron springs could mitigate the feeling akin to seasickness inspired by hanging from the leather straps on top of the coach. Robinson, *British Post Office*, 128–129; Ralph Straus, *Carriages and Coaches: Their History and Their Evolution* (London: Martin Secker, 1912), 171–264.

149. A. M. Ogilvie, "The Rise of the English Post Office," *Economic Journal* 3:11 (1893): 443–457; A. Newlin, "An Exhibition of Carriage Designs," *Metropolitan Museum of Art Bulletin* 35:10 (1940): 185–191; Robinson, *British Post Office*, 230; H. J. Dyos and D. H. Aldcroft, *British Transport: An Economic Survey from the Seventeenth Century to the Twentieth* (Leicester: Leicester University Press, 1969); A. H. Nelson, "Six-Wheeled Carts: An Underview," *Technology and Culture* 13:3 (1972): 391–416; G. Turnbull, "State Regulation in the Eighteenth-Century English Economy: Another Look at Carriers' Rates," *Journal of Transport History* 6 (1985): 18–36; D. Gerhold, *Road Transport before the Railways: Russell's London Flying Waggons* (Cambridge: Cambridge University Press, 1993); Dorian Gerhold and Theodore Cardwell Barker, *The Rise and Rise of Road Transport, 1700–1990* (Cambridge: Cambridge University Press, 1995); A. Puetz, "Design Instruction for Artisans in Eighteenth-Century Britain," *Journal of Design History* 12:3 (1999): 217–239; D. King-Hele, "Erasmus Darwin's Improved Design for Steering Carriages—and Cars," *Notes and Records* 56:1 (2002): 41–62; J. Herson, "Estimating Traffic," *Journal of Transport History* 23:2 (September 2002): 113–146.

150. O'Flanagan, *Impressions at Home and Abroad*, 85.

151. Henry Mackenzie, *The Works of Henry Mackenzie: With a Sketch of the Author's Life* (London: T. Cadell, 1822), 62.

152. Robert Southey, *Letters from England*, 2 vols. (London: Printed for Longman, 1808), 1:354–358.

153. "Travelling," *Visitor, or Monthly Instructor*, 1838, 310.

154. David Hughson, *London: Being an Accurate History and Description of the British Metropolis* (London: J. Stratford, 1807), 454; "Night Walker," 508.

155. "The Steam Navigation Inquiry," *Mechanics Magazine*, January 7, 1832, 263.

156. Southey, *Letters from England*, 1:354.

157. Richard Rush, *A Residence at the Court of London* (London: R. Bentley, 1833), 49.

158. James Grant, *The Great Metropolis* (New York: Theodore Foster, 1837), 12.

159. "Henry Wombill Indicted for Killing and Slaying Esther Hoe," *Central Criminal Court*, 1840, 599–601.

160. Common travelers who slept in barns did not fraternize with those who slept at inns. As Charles Moritz recorded of his seven weeks' visit to England in 1782, his

decision to journey on foot was met with surprise by innkeepers, who were unaccustomed to receiving visitors who had not come by coach, and the manner in which he traveled "exposed him to the scorn of a people whom he wished to respect." Inns were a scene for encountering a delimited and respectable metropolitanism in a safe environment purged of its potential diversity. Quoted in Robert Southey, "On the Accounts of England by Foreign Travelers," in *Essays, Moral and Political*, 2 vols. (London: John Murray, 1832), 1:263.

161. The *Gentleman's Magazine* could point out to antiquarians as recently as 1806 that inns called "the Bowling Green" and "the Skittle Ground" had once served those purposes: "Remarks on the Signs of Inns, &c," *Gentleman's Magazine*, April 1818, 309. William Hogarth's *Election* series depicted the remnant of this sociability, the hustings conducted amid rioting, courtship, trade, and drunkenness at a local inn. During the civil and religious frays of the sixteenth and seventeenth centuries, inns served as courts of law and prisons. Quaker Elias Osborn's congregation met at an inn at Ilchester. Prominent seventeenth-century Quakers Gilbert Latey and David Hall were children of innkeepers.

162. Obituary, "Paul Sandby," *Universal Magazine* 12 (1809): 512.

163. John Hassell, *Tour of the Grand Junction Illustrated with a Series of Engravings* (London: Printed for J. Hassell, 1819).

164. D. E. Allen, *The Naturalist in Britain: A Social History* (Princeton, N.J.: Princeton University Press, 1994), esp. chap. 1; B. Silver, "William Bartram's and Other Eighteenth-Century Accounts of Nature," *Journal of the History of Ideas* 39:4 (1978): 597–614.

165. Elsewhere, inns began to cater in appearance and entertainment to this class with its scientific interests in the culture of collecting. Some inns sported their own museums, while others housed traveling exhibits. Inns accommodated local debating unions, philosophical societies, and geologic clubs and doubled as printing offices. William Clarke, *Publican and Innkeeper's Practical Guide* (London: Sherwood, 1829); F. W. Hackwood, *Inns, Ales, and Drinking Customs of Old England* (London: T. F. Unwin, 1909); A. E. Richardson, *The English Inn, Past and Present* (London: J. B. Lippincott Company, 1926), 91–137; J. Fothergill, *An Innkeeper's Diary* (London: Chatto and Windus, 1931); G. Hogg, *The English Country Inn* (London: Batsford, 1974); Margetson, *Journey by Stages*; Alan Milner Everitt, "The English Urban Inn, 1560–1760," chap. 8 in *Landscape and Community in England* (London: Hambledon Press, 1985), 155–200.

166. Famous actors James Biddles, James Cowles, John Collins, William Dowton, Edward Cape Everard, John Lowin, Elizabeth Fitzhenry, and George Parker were children of innkeepers, raised among the flux of personalities and interactions that typified inn society. Military cartographers and sons of innkeepers composed two major demographics of the first generation of

landscape painters. The landscape painters George Paul Chalmers, Thomas and William Daniell, Julius Caesar Ibbetson, Benjamin West, and William Woolett and the prominent engraver Thomas Bewick were sons of innkeepers. Scientists included Jacob Bobart, Charles Jenner, and Robert Uvedale, botanists; James Douglas and Edwin Witchell, geologists; Elizabeth Gray, the fossil collector; Charles William Peach, the naturalist; Henry Piddington, the meteorologist; Jesse Ramsden, the maker of scientific instruments; and Henry Faulds, the developer of fingerprinting. Virtually every profession associated with the Enlightenment depended on travel. At any inn across England, some part of the usual conversation was likely to involve the kinds of knowledge required for the obsessive cataloging of the environment.

167. S. H. Spiker, *Travels through England, Wales, and Scotland, in the Year 1816* (London: Printed for Lackington, Hughes, Mavor, and Jones, 1820), 223.

168. Obituary, "Thomas Salter," *New Monthly Magazine*, July 1, 1816, 84.

169. Edward Wedlake Brayley, *Londiniana; Or, Reminiscences of the British Metropolis* (London: Hurst, Chance, and Co., 1829), 270.

170. John Gamble, *Sketches of History, Politics, and Manners, in Dublin* (London: Baldwin, Craddock, and Joy, 1826), 5.

171. H. W. Hart, *Sherman and the Bull and Mouth* (Leicester: Leicester University Press, 1961); David Watkin, *Grand Hotel: The Golden Age of Palace Hotels; An Architectural and Social History* (London: Vendome Press, 1984).

172. Richardson, *English Inn*, 112.

173. "The Bachelor's Vade-Mecum," *New Monthly Magazine* 23:92 (August 1828): 174.

174. Frederick Reynolds, *The Life and Times of Frederick Reynolds* (London: Henry Colburn, 1826), 94.

175. O'Flanagan, *Impressions at Home and Abroad*, 83.

176. Reynolds, *Life and Times of Frederick Reynolds*, 94.

177. Catherine Delano-Smith and Roger Kain, *English Maps: A History* (Toronto: University of Toronto Press, 1999).

178. Cf. Thomas Badeslade, *A Compleat Sett of Mapps of England and Wales*, An instance of the book as luxury object, the volume contains forty-nine full-color maps of England's principal post roads, direct roads, counties, and rivers, which were hand painted on thin sheets of smooth vellum by Thomas Badeslade himself.

179. Matthew Simons, *A Direction for the English Traviller* (London: Printed by John Garrett, [1677?]), measures 13 cm and includes maps that fold out to four times that size: still a small but unified image of the network. John Ogilby's strip maps contain a single folded leaf plate of the road network bound facing the title page. The earliest is Ogilby's *Britannia* of 1675, measuring 35 cm. It

was gradually reduced, and by 1681 John Ogilby and William Morgan produced *The Traveller's Pocket-Book*, with 278 pages, measuring 13 cm. John Ogilby's strip maps first included a single folded-leaf plate of the road network bound facing the title page, but these were eliminated after 1720. The 1716 edition of Ogilby's work displayed a primitive, four-route image of the road network; the 1724 edition dispensed with an image of the network altogether.

180. Catherine Delano-Smith has emphasized the difference between seventeenth-century route making, writing out in advance the names of places one would pass on an itinerary without knowledge of the road, and way finding, which could be done on the road. Way finding with a map required a mental map of what the area traversed looked like, and way finding at a distance required an image of the island as a nation connected by certain roads. Catherine Delano-Smith, "Milieus of Mobility: Itineraries, Route Maps and Road Maps," in James Akerman, *Cartographies of Travel and Navigation* (Chicago: University of Chicago Press, 2006), chap. 2.

181. *Bowles's New Traveller's Guide* (1777) is an early example of the genre. A single 48¥36 cm sheet, it showed the direct roads and crossroads of England and Wales. It was reprinted as *Bowles's New Pocket Guide* in 1780. We have no record of any further editions. Daniel Paterson began printing his itinerary, *A New and Accurate Description of All the Direct and Principal Cross Roads*, in 1781. He reprinted it in 1784, 1785, 1786, 1789, 1792, 1794, 1796, 1808, and 1811. Paterson's *Description* and Cary's *Travellers Companion* (1790 with multiple editions) were essentially more exact redactions of Ogilby's strip-map road books that first appeared in 1675.

182. An advertisement printed in the back of Thomas Bowles, *Post Chaise Companion* (London: Thomas Bowles, 1782), lists the new pocket map of Scotland, according to the survey of James Dorret, geographer, for two shillings. Cary's map in 1798 was selling for seven shillings. This may indicate the comparative value of the carefully surveyed Cary maps.

183. Farington's illustrated travels include Joseph Farington, *Views of the Lakes, etc. in Cumberland and Westmorland* (London: Published by William Byrne, 1789). See also Daniel and Samuel Lysons, *Magna Britannia* (London: Printed for T. Cadell and W. Davies, 1806–1822); Thomas Pennant, *The History and Antiquities of London* (London: Printed for Edward Jeffery by B. McMillan, 1814).

184. By the 1780s, the theory of the picturesque reached a middling audience who learned how to read the landscape from the writings of William Gilpin. As early as the 1780s, literary tourbooks guided travelers to landmarks of Shakespeare's Britain, and later to landmarks of Robert Burns and Sir Walter Scott. Ian Ousby, *The Englishman's England: Taste, Travel, and the Rise of Tourism*

(Cambridge: Cambridge University Press, 1990), 168–169; Nicola J. Watson, *The Literary Tourist: Readers and Places in Romantic and Victorian Britain* (Basingstoke: Palgrave Macmillan, 2006). Watson gives these early examples: William Henry Ireland, *Picturesque Views on the Upper, or Warwickshire Avon* (1795); Charles Mackenzie, *Interesting and Remarkable Places* (1832); Thomas Dibdin, *A Bibliographical, Antiquarian, and Picturesque Tour in the Northern Counties of England and Scotland* (1838); *The Land of Burns* (1840); William Howitt, *Visits to Remarkable Places* (1840, 1842); and *Homes and Haunts of the Most Eminent British Poets* (1847). She associates the rise of this genre with the longer tradition of visiting the graves of famous poets, well in swing by the end of the century. G. Adams, *Travelers and Travel Liars, 1660–1800* (Berkeley: University of California Press, 1962), 90–91. Examples of natural science as the excuse for tourism are John Phillips, *Palaeozoic Fossils of Cornwall* (London: Longman, Brown, Green, and Longmans, 1841); William Turton, *A Manual of the Land and Fresh-water Shells of the British Islands* (London: Longman, Orme, Brown, 1840); George Nicholson, *The Cambrian Traveller's Guide* (Stourport: George Nicholson, 1808); Thomas Walford, *The Scientific Tourist through England, Wales and Scotland* (Printed for J. Booth, 1818); and Thomas Walford, *The Scientific Tourist through Ireland* (London: J. Booth, 1818).

185. Southey, *Letters from England*, 1:359.
186. Gamble, *Sketches of History, Politics, and Manners*, 39–40.
187. "Stage Coach Travelling," *Olio, or Museum of Entertainment* 11 (1833): 240
188. S. S., *A Visit to Edinburgh* (Edinburgh: Fairburg and Anderson, 1818), 8–9.
189. Robert Heron, *The Comforts of Human Life; or, Smiles and Laughter of Charles Chearful* (London: J. and W. Smith, 1807), 50.
190. Thomas Staunton St. Clair, *A Soldier's Recollections of the West Indies and America* (London: R. Bentley, 1834), 261.
191. William Dunlap, *Memoirs of the Life of George Frederick Cooke, Esquire* (New York: D. Longworth, 1813), 106.
192. "Travelling," *Visitor, or Monthly Instructor*, 1838, 310.
193. Gamble, *Sketches of History, Politics, and Manners*, 4.
194. "Stage Coach Adventures," *Table Book* 1 (1827): 263.
195. "The Stage-coach," *Edinburgh Magazine*, September 1824, 349.
196. *A Manual of Politeness: Comprising the Principles of Etiquette* (London: H. G. Bohn, [1837] 1850), 146.
197. Mackenzie, *Works of Henry Mackenzie*; "The Life of a Sub-editor," *Metropolitan Magazine* 12 (January–April 1835), 95–104; "Stage Coach Adventures"; Heron, *Comforts of Human Life*, 126–142.
198. Heron, *Comforts of Human Life*, 50.
199. "Life of a Sub-editor," 95.

200. Southey, *Letters from England*, 1:354–358. For other examples, see Gamble, *Sketches of History, Politics, and Manners*, 3; Caroline Fry Wilson, *The Listener*, 2 vols. (London: Latimer and Co., 1832), 2:175; "Stage-coach," 354; and Anne Jackson Matthews, *A Continuation of the Memoirs of Charles Mathews, Comedian*, 2 vols. (Philadelphia: Lea and Blanchard, 1830), 1:202, 237.
201. "Stage-coach"; "Stage Coach Adventures"; John Farrar, *The Young Lady's Friend* (Boston: American Stationers' Company, 1836); *A Manual of Politeness: Comprising the Principles of Etiquette* (London: H. G. Bohn, [1829] 1850), 146.
202. Farrar, *Young Lady's Friend*, 401.
203. Joseph Addison and Richard Steele, *The Spectator* 8:631 (December 10, 1728), 280–282.
204. Farrar, *Young Lady's Friend*, 401.
205. Mackenzie, *Works of Henry Mackenzie*, 62.
206. Richard Gooch, *Facetiae Cantabrigienses* (London: Baily, 1824), 134.
207. Mackenzie, *Works of Henry Mackenzie*, 64.
208. "Small Talk," *Blackwood's Edinburgh Magazine*, September 1819, 686.
209. *The Young Gentleman and Lady's Monitor, and English Teacher's Assistant* (London: S. Green, 1802), 175.
210. "A Few Thoughts on Small-Talk," *New Monthly Magazine* 5 (January–June 1823): 217.
211. Gooch, *Facetiae Cantabrigienses*, 134.
212. "Few Thoughts on Small-Talk."
213. Farrar, *Young Lady's Friend*, 401.
214. David Ashforth, "Settlement and Removal in Urban Areas: Bradford, 1834–71," in Michael E. Rose, ed., *The Poor and the City: The English Poor Law in Its Urban Context, 1834–1914* (New York: Leicester University Press, 1985).
215. Quoted in *The Roads and Railroads, Vehicles, and Modes of Travelling, of Ancient and Modern Countries* (London: John W. Park, 1839), 3.
216. *Metropolitan* 1 (May–August 1831): 143.
217. George Richardson Porter, *The Progress of the Nation* (London: Knight, 1836), 10, quoting Mr. Loch, who testified before the 1835 Select Committee on Public Works in Ireland.
218. *Roads and Railroads*, iii.
219. W. M. Gillespie, *A Manual of the Principles and Practice of Road-Making*, 2nd ed. (New York: A. S. Barnes and Co., 1848), 21.
220. "Social Effects of Railways," *Chambers's Edinburgh Journal* 38 (September 21, 1844): 1.
221. Porter, *Progress of the Nation*, 11.
222. *Roads and Railroads*, 65.
223. Porter, *Progress of the Nation*, 11.

224. John Richard Digby Beste, *Odious Comparisons; or, The Cosmopolite in England*, 2 vols. (London: Saunders and Otley, 1839), 2:1–2.
225. Michael Angelo Garvey, *The Silent Revolution; or, The Future Effects of Steam and Electricity upon the Condition of Mankind* (London: William and Frederick G. Cash, 1852), 98–99.
226. "Once upon a Time," *Blackwood's Edinburgh Magazine*, June 1855, 697.
227. Garvey, *Silent Revolution*, 55, 66.
228. Thomas Babington Macaulay, *History of England* (London: Longman, Brown, Green, and Longmans, 1849), 1:296, 332, 374.
229. Ibid., 1:370–371.
230. Ibid., 1:371.
231. This quotation is found only in the Boston edition of the 1849 printing and later editions printed in Britain. Thomas Babington Macaulay, *History of England* (Boston: Phillips, Samson, and Company, 1849), 1:345. All other notes reference the original London printing.

Conclusion

1. Michael Angelo Garvey, *The Silent Revolution; or, The Future Effects of Steam and Electricity upon the Condition of Mankind* (London: William and Frederick G. Cash, 1852), chap. 14.
2. John Stuart Mill, *On Liberty* (London: J. W. Parker, 1859), 198.
3. Miles Ogborn, "Local Power and State Regulation in Nineteenth Century Britain," *Transactions of the Institute of British Geographers* 17 (1992): 215–226; R. Lambert, "Central and Local Relations in Mid-Victorian England: The Local Government Act Office, 1858–1971," *Victorian Studies* 6 (1962): 121–150; R. Gutchen, "Local Improvements and Centralisation in Nineteenth Century England," *Historical Journal* 4 (1961): 85–96.
4. "Road Reform," *Journal of Agriculture* (January 1857), 681 25 & 26 Vict., c. 61 (Act for the Better Management of the Highways of England, 1863–4); "History of Highways," *Edinburgh Review*, April 1864, 368.
5. Arnold Toynbee and Benjamin Jowett, *Lectures on the Industrial Revolution of the 18th Century in England* (London: Rivingtons, 1887), 91; Ellen Louise Osgood, *A History of Industry* (Boston: Ginn and Company, 1921), 287–288; Lilian Charlotte Anne Knowles, *The Industrial and Commercial Revolutions in Great Britain during the Nineteenth Century* (New York: E. P. Dutton & Co., 1921), 410; Sidney Webb and Beatrice Webb, *English Local Government: The Story of the King's Highway* (London: Longmans, Green, 1913).
6. Select Committee on Turnpike Trusts and Tolls, *Report*, 1836 (647).
7. Ibid.; "Turnpike Trusts," *Jurist*, July 15, 1864, 265–266.

8. 14 & 15 Vict., c. 37; House of Commons, "Report of the Secretary of State for the Home Department Relative to Turnpike Trusts, Reporting under 3 & 4 Wm. IV c. 80," *Parliamentary Papers*, 1852–1853; "Turnpike Trusts," 265–266.
9. *Parliamentary Debates*, April 6, 1821, 69.
10. Coms. of Metropolitan Turnpike Roads North of Thames, *Second Report*, 1828 (331); 7 Geo. 4, c. 142 (Metropolis Roads Act, 1827); Coms. Of Metropolitan Turnpike Roads North of Thames, *Third Report*, 1829 (219); *Fourth Report*, 1830 (362); *Fifth Report*, 1831 (41); *Sixth Report*, 1831–1832 (449); *Seventh Report*, 1834 (237); *Eighth Report*, 1834 (238); *Ninth Report*, 1835 (208); *Tenth Report*, 1836 (616); *Eleventh Report*, 1837 (305); *Twelfth Report*, 1837–1838 (387); 9 & 10 Vict., cc. 59, 83 (Metropolitan Improvements Act, 1846–1847); 21 & 22 Vict., c. 66 (Chelsea Bridge Act, 1858–1859).
11. Mark Searle, *Turnpikes and Toll-bars* (London: Hutchinson and Co., 1930), 662–663; *Times*, July 21, 1825, cited in Searle, 662.
12. Searle reprints the *Times* reports on the bill for April 25 and May 2, 1825. Searle, *Turnpikes and Toll-bars*, 222–223.
13. Searle, *Turnpikes and Toll-bars*, 222–224; *Hansard's Parliamentary Debates* (March 12, 1830); *Bell's Life in London*, January 10, 1830.
14. In 1857, the Toll Reform Association, chaired by Sir Joseph Paxton, member of Parliament and architect, sent a petition to Parliament to abolish all tollgates within six miles of Charing Cross. Its petition also hoped to overturn the Metropolis Turnpike Commission, a self-elected, closed-door body, and to turn its functions over to the open-door Metropolitan Board of Work (hitherto in charge of the sewers).
15. Select Committee on the Expediency and Practicability of Abolishing Turnpike Trusts, *Report*, 1864 (383).
16. Great Britain and Alexander Glen, *The Acts for the Better Management of the Highways in England, 1862–1864* (London: Knight, [1862] 1875); 28 & 29 Vict., c. 107 (Act to Continue Certain Turnpike Acts, 1865); 3 Geo. 4, c. 126, sects. 118 and 124 (1824); 29 & 30 Vict., c. 105 (August 1866); 30 & 31 Vict., c. 121 (1867); 31 & 32 Vict., c. 99 (1868); 32 & 33 Vict., c. 90 (1869); 34 & 35 Vict., c. 115 (1871); 35 & 36 Vict., c. 85 (1872); 36 & 37 Vict., c. 90 (1873); 37 & 38 Vict., c. 95 (1874); 35 & 36 Vict., c. 79 (Local Government Act, 1872); "History of Highways," 366; 7 & 8 Vict., c. 91. (1846).
17. "History of Highways," 367. As Mr. Wrightson observed in the 1864 report, "The through traffic has been diverted by railroads; the roads have become a local burthen, and the trustees have been driven to the multiplication of bars." Quoted in "Turnpike Trusts," *Jurist*, July 15, 1864, 265–266.
18. "Road Reform," 598.
19. William Pagan, *Road Reform: A Plan for Abolishing Turnpike Tolls and Statute Labour* (Edinburgh: Blackwood and Sons, 1845).

20. Interview, Sir James McAdam, June 20, 1836, Select Committee on Turnpike Trusts and Tolls, *Report*, 1836 (647), 125; *The Statistical Account of Lanarkshire* (Edinburgh: W. Blackwood, 1841), 700; Jenner Marshall, *Memorials of Westcott Barton, in the County of Oxford* (London: J. R. Toulmin Smith, 1870), 52; Edmund Leahy, *A Practical Treatise on the Making and Repair of Roads* (London: J. Weale, 1844), 22; "Road Reform," 681.
21. The average amount of tolls gathered in the same period was £1,490,517, the majority of which was spent on basic repairs and management, at £1,122,000, and interest on the tolls, at £300,000 a year. Royal Commission for Inquiry into the State of Roads in England and Wales, *Report*, 1840 (256).
22. Ibid. By 1850 tolls would have begun to decline on their own.
23. William B. Chorley, *A Handybook of Social Intercourse* (London: Longman, Green, Longman, and Roberts, 1862), 40.
24. George Cornewall Lewis argued that "placing the highways under the authorities of the vestries was a mode of proceeding repugnant to all sound principles of management; and that that species of road could never be placed upon a satisfactory footing till the mode of its management could be totally changed." "Highways," *Hansard's Parliamentary Debates* (February 13, 1850), 748.
25. "History of Highways," 186.
26. "The Central Chamber of Agriculture," *Farmer's Magazine* (March 1867): 256.
27. William Carnie, *Reporting Reminiscences* (Aberdeen: Aberdeen University Press, 1904), 196.
28. Homersham Cox, *The Institutions of the English Government* (London: H. Sweet, 1863), 189.
29. "Central Farmer's Club," *British Farmer's Magazine* 41:22 (1867): 233–239.
30. Jenkinson called it an outrage that "the owners of £800,000,000 of funded property should be exempt from contributing one penny towards the various local burdens." He pointed out that the "maintenance of the turnpikes and highways" "pressed exclusively upon the agricultural class." "Central Chamber of Agriculture: The Education of the Agricultural Laborer," *Farmer's Magazine* 2 (1868): 428.
31. "The New Highway Act," *Farmer's Magazine* (May 1868): 381.
32. 51 & 52 Vict., c. 41.
33. "Report from the Select Committee on Intended Improvements in the Post-Office," Post Office Archive, 91/3, 1815.
34. David P. Jordan, *Transforming Paris: The Life and Labors of Baron Haussmann* (New York: Free Press, 1995); Dirk Schubert and Anthony Sutcliffe, "The 'Haussmannization' of London? The Planning and Construction of Kingsway-Aldwych, 1889–1935," *Planning Perspectives* 11, no. 2 (1996): 115.
35. Boyd Hilton, *The Age of Atonement: The Influence of Evangelicalism on Social and Economic Thought, 1785–1865* (Oxford: Clarendon Press, 1988); Christo-

pher Hamlin, *Public Health and Social Justice in the Age of Chadwick: Britain, 1800–1854* (Cambridge: Cambridge University Press, 1998).
36. Pamela K. Gilbert, *Mapping the Victorian Social Body* (Albany: State University of New York Press, 2004); Mary Burgan, "Mapping Contagion in Victorian London: Disease in the East End," in Debra N. Mancoff and D. J. Trela, eds., *Victorian Urban Settings: Essays on the Nineteenth-Century City and Its Contexts* (New York: Garland, 1996), 43–56; Christopher Otter, "Cleansing and Clarifying: Technology and Perception in Nineteenth-Century London," *Journal of British Studies* 43:1 (2004): 40–64; Christopher Otter, *The Victorian Eye: A Political History of Light and Vision in Britain, 1800–1910* (Chicago: University of Chicago Press, 2008).
37. A. J. Christopher, "Roots of Urban Segregation: South Africa at Union, 1910," *Journal of Historical Geography* 14:2 (April 1988): 151–169.
38. D. Maudlin, "Robert Mylne, Thomas Telford and the Architecture of Improvement: The Planned Villages of the British Fisheries Society, 1786–1817," *Urban History* 34:3 (2007): 453–480.
39. W. Ashworth, "British Industrial Villages in the Nineteenth Century," *Economic History Review*, n.s., 3:3 (1951): 378–387.
40. Margaret Crawford, "The 'New' Company Town," *Perspecta* 30 (1999): 48–57.
41. James W. Loewen, *Sundown Towns* (New York: New Press, 2005); J. N. Day, "A Covenant of Color: Race and Social Power in Brooklyn," *Journal of Social History* 35 (2002): 719–721; K. F. Gotham, "Urban Space, Restrictive Covenants and the Origins of Racial Residential Segregation in a US City, 1900–50," *International Journal of Urban and Regional Research* 24:3 (2000): 616–633; Michael Jones-Correa, "The Origins and Diffusion of Racial Restrictive Covenants," *Political Science Quarterly* 115:4 (Winter 2000–2001): 541–568; Robert C. Weaver, *The Negro Ghetto* (New York: Harcourt, Brace, 1948); Herman H. Long and Charles S. Johnson, *People vs. Property* (Nashville, Tenn.: Fisk University Press, 1947); Charles Abrams, *Forbidden Neighbors* (New York: Harper and Brothers, 1955).
42. Stephen Macedo and Christopher F. Karpowitz, "The Local Roots of American Inequality," *PS: Political Science and Politics* 39:1 (2006): 59–64; Thomas Osborne and Nikolas Rose, "Governing Cities: Notes on the Spatialisation of Virtue," *Environment and Planning* 17:1 (1999): 737–760; Simon F. Parker, "Community, Social Identity and the Structuration of Power in the Contemporary European City," *City* 5:2 (2001): 281–309.
43. Alison Ravetz, *Council Housing and Culture: The History of a Social Experiment* (New York: Routledge, 2001).
44. Dick Hebdige, *Hiding in the Light: On Images and Things* (London: Routledge, 1988).
45. David Harvey, *Paris, Capital of Modernity* (New York: Routledge, 2003).

46. Thomas J. Sugrue, *The Origins of the Urban Crisis: Race and Inequality in Postwar Detroit* (Princeton, N.J.: Princeton University Press, 1996). See also M. Fullilove, *Root Shock: How Tearing Up City Neighborhoods Hurts America, and What We Can Do about It* (New York: Ballantine Books, 2005).
47. Robert Goodman, *After the Planners* (New York: Simon and Schuster, 1972), 21.
48. Joe Moran, "The Future of Roads," *Soundings* 44 (March 25, 2010): 107–116; "Mapping Out the Future of Transport," BBC, January 27, 2006; Ben Webster and Michael Evans, "Radical Dreams for the Future of Transport Haunted by Past Failures," *Times*, June 6, 2005, http://www.timesonline.co.uk/tol/news/uk/article530470.ece.
49. John Seely Brown, *The Social Life of Information* (Boston: Harvard Business School Press, 2000), xviii.
50. Fred Turner, *From Counterculture to Cyberculture: Stewart Brand, the Whole Earth Network, and the Rise of Digital Utopianism* (Chicago: University of Chicago Press, 2006).
51. "IBM Sustainable Cities—Ideas—United States," April 8, 2010, http://www.ibm.com/smarterplanet/us/en/sustainable_cities/ideas/.
52. John Geraci, "DIYCity," http://www.diycity.org; John Geraci, "The Future of Our Cities: Open, Crowdsourced, and Participatory—O'Reilly Radar," *O'Reilly Radar*, April 6, 2009, http://radar.oreilly.com/2009/04/the-future-of-our-cities-open.html.
53. Gretchen Livingston and Susannah Fox, *Latinos Online: Narrowing the Gap—Pew Research Center*, Pew Internet and American Life Project (December 22, 2009), http://pewresearch.org/pubs/1448/latinos-internet–usage-increase-2006-2008.
54. J. Horrigan and K. Murray, *Home Broadband Adoption*, Pew Internet and American life project (2006), http://www.pewinternet.org/Reports/2006/Home-Broadband-Adoption-in-Rural-America.aspx; P. Stenberg, Mitch Morehart, Stephen Vogel, John Cromartie, Vince Breneman, and Dennis Brown, *Broadband Internet's Value for Rural America* (Washington, D.C.: U.S. Department of Agriculture, Economic Research Service, 2009).

ACKNOWLEDGMENTS

This project received substantial support from the Harvard Society of Fellows; the History Department at the University of Chicago; the Huntington Library; the Yale Center for British Art; the Paul Mellon Centre for Art and Architecture; the Dumbarton Oaks Landscape Architecture Library; Pembroke College, Cambridge; and the Anglo-California Society, the Institute for Humane Studies, and the Center for British Studies and Department of History at the University of California, Berkeley. Librarians Sally North at the Royal Institute of British Architects and Elisabeth Fairman at the Yale Center for British Art led me to treasures underground.

Particular thanks are due to Michael Mason, my editor, and Kathleen McDermott, who worked with me at Harvard University Press. Simon Murray in London, Woytek Rakowitz and Peter Cavanagh in Cambridge, and Alexander Cockburn in Humboldt gave me warm places to stay and rich conversation to digest during the initial phases of research for this project. Investigating the questions of this book was indeed possible only with guides willing to help me explore fields about which I was profoundly naïve ten years ago: landscape studies and British history. Philip Howell, Paul Groth, and Tom Laqueur deserve deep thanks, much more than I was able to render them at the time. Various drafts were read by Peter Onuf, David Henkin, Tim Hitchcock, Fredrick Albritton Jonnson, Adrian Johns, Tim Sparrow, Bonnie Wheeler, Jeremy du Quesnay Adams, Nicholas Hoover Wilson, Michael Shapiro, Catherine Karnitas, Penny Ismay, Caroline Shaw, Desmond FitzGibbon, John Gillis, Jonathan Grossman, Elizabeth Honig, David Lieberman, Simon Gunn, and especially Rosalind Williams, from whose patient input and substantial commentary the project benefited enormously. Other interlocutors, such as Robin Einhorn, Michael Munger, and David Schmidtz, kept me face to face with the political struggles between capital and periphery, and this book is in many respects an attempt to explore the difficult questions they posed. Chris Robling and Jerry Michalski pushed me to connect eighteenth-century problems to contemporary ones. Conversations with Jon Parry, Martin Daunton, Peter Mandler, Christina Gillis, Leon Litwack, Susanna Barrows, and Richard Candida Smith also pointed me to crucial thresholds along the way.

ACKNOWLEDGMENTS

There are many others, of course, who taught me, talked to me, and sustained the energies behind this project, but this book is dedicated to two mentors whose loyalty to ideas and craft carried the project past every adversity, including those posed by my own stubbornness and ignorance: John Stilgoe, who taught me to observe, and James Vernon, who taught me how to write.

INDEX

Access, to roads, 133–138
Accidents, government provisions against, 112–113

Board of Agriculture, 91, 101–102, 104, 108
Bodies in pain, as spectacle, 112–113
Bridges, 5, 12, 16
Britain, 14–15, 17, 22, 29; roads integrating, 106–120, 194
Broadband, as infrastructure, 211–212
Built environment, 4, 5, 75, 79–80, 122, 124–127
Bureaucracy, 21, 29, 37, 47, 80–88, 120, 125–127, 198
Bureaucrats, 131, 138–146
Burt, Edmund, 30–31, 40–41

Canals, 6–8, 14–15, 27–28, 36, 52
Cartography, 18th-century innovations in, 32–36. *See also* Maps
Centralization: of roads, 23, 48, 90, 145; of information, 49; under parliamentary building, 80–82; envisioned in 1690s, 82–88; financial efficiency of, 94–95; local corruption and, 96–99; economic participation by former colonies and, 100; as measure against interestedness, 119–120; road repairs and, 121; of turnpike trusts, 123–127; opposition to, 130–131; inefficiency of, 138–146; as poison, 144; as source of bad roads, 145; of infrastructure, 148
Centralized board of roads: civil engineers advocating, 93; plans for, 135

Centralizers, 18; local government and, 104–106; single authority proposed by, 119; under attack, 140–146
Chartists, opposition to centralization, 130, 145, 150
Circulation, roads and, 108, 110
Cities, cosmopolitan ideas in, 20–21, 23, 182
Civil engineers. *See* Engineers, civil
Class, 2–3, 21–22. *See also* Labor; Middle class, creation of; Poverty
Coasts, 13, 34
Cobbett, William, 130, 133–134, 142, 199, 209
Colonial Scotland and Ireland, roads lacking in, 87
Colonies, collapse of, 88
Commons, roads as, 47, 199
Community: as alternative to nation, 22; infrastructure, 24; government and, 130–133; among travelers, 154–155, 174–178
Compass, introduction of, 33
Competition: between infrastructure providers, 8, 86; local governments stifling, 90. *See also* Free market
Condorcet, Nicolas de, 7, 89
Congestion charges, 209
Construction. *See* Engineering
Consumption, 155
Copyright, 4
Corps des Ponts et Chaussées, 14, 32, 90
Corruption: of local government, 86; of surveyors, 141
Cosmopolitanism: of city, 20–21, 23, 182; of road, 21

Defoe, Daniel, 7, 13, 82–88
Development: government influence on, 16; centralization promoting, 94–95; regional, 110
Diagrams: as a political tool, 27–28; argumentation and, 66
Dublin, London connected to, 80

Economic participation, 9, 199; centralized highways promising, 100; tolls and, 130–132
Edgeworth, Richard Lovell, 51
Edinburgh, London connected to, 80
Eminent domain, 3, 23, 206; expansion of, 78; private property abrogated by, 78; limited, 86; local corruption and, 86; radicals attacking, 143
Engineering: military, 13; parliamentary and military, 27; birth of civil engineering, 50; lobbying transforming, 51–53; mounting expense as result of, 76. *See also* Road construction
Engineers, Army Corps of, 18
Engineers, civil: hero status of, 4; lobbying by, 27, 46–47, 49, 51–53; birth of, 46–47, 50; self-taught, 49; general rules dominating, 51–56; as authority figures, 73; local context disregarded by, 78; local surveyors compared to, 104; mounting expenses of, 139, 141; skepticism toward, 142. *See also* Corps des Ponts et Chaussées
Engineers, Dutch, 7
Engineers, military, 14, 16; at Ordnance Survey, 14; new geometry used by, 34–35; strict organization of, 45–46
Etiquette manuals, 187–188
Evictions, to make room for roads, 78
Expense: of expert rule, 5, 12; of connecting periphery, 18; of modern infrastructure, 54, 84; rigor of calculating, 71; of bridges, 76; of maintenance, 76–77; unforeseen, 76–77; unprecedented, 84
Expert opinion, 48
Experts: rule of, 4; cost of 5, 12; manuals and forms used by, 15; in parliamentary reports, 49; parliamentary procedure dominated by, 52–53, 61–62; excessive use of paper by, 62–63; travel expenses and, 66, 68; as surgeons of estimating expense, 71; authority of, 73; consensus, top-down governance and, 76; financial efficiency of, 94–95; corruption and, 96–99, 141; as "secret boards," 131, 144; inefficiency of, 138–146; skepticism toward, 142; powers curtailed, 146–152; in public health, poverty, and urban planning, 206; jurisdiction over public space, 207–208. *See also* Bureaucracy; Government
Expert witnesses, 47

Finance, as political tool, 27–28. *See also* Expense
Forced purchase. *See* Eminent domain
Free market: canal and rail construction influenced by, 15; economy driven by, 15–16; in turnpike trusts, 17; triumph of, 28; crooked roads produced by, 37; negotiation for land and, 39; roads and, 90; as determinant of investment, 140; users paying for roads, 143–144. *See also* Competition; Localism
Free roads, radical movement toward, 145
Free trade, 71, 148
Friction, expense of, 68

Geographical periphery, paying for roads, 201
Geography: government shaped by, 18; political parties and, 18; single, emergence of, 44; roads to nowhere and, 61; roads redistributing trade over, 80–81; tendency toward local monopoly, 90, 95; region as political lobby, 109; inequality and, 204; of U.S. rust belt, 210–211; political tensions across, 212
Geology: in traditional road making, 44; standardized, 69
Globalization, British infrastructure driving, 15
Government: infrastructure design and, 4; proprietary infrastructure protected by, 8; expansion of, 13, 15, 47; taxes and, 16; infrastructure no longer promoted by, 18; socialist, 18–19; everyday behavior monitored by, 19; participatory, 23; as builder of highways, 46–53, 120; parliamentary decision making and, 46–78; Post Office and, 47; cheap, 71, 80; top-down format, 76; limited role over infrastructure, 85–86; infrastructure competition managed

INDEX

by, 86; efficiency of free market and, 143; open participation and, 144; requirements of accountability and, 145; suspicion of, 146–152; new age of travel prompted by, 154; speed regulated by, 180–181. *See also* Bureaucracy; Centralization; Eminent domain; Experts

Governmentality. *See* Labor; License plates; Snowstorms; Traffic

Government contracts, for carriage manufacture, 180

Government reform, age of, 80–81

Guidebooks, 2, 183–187

Gypsies, 172–173

Hierarchy, in labor organization, 45–46

Highland Roads and Bridges, 93–94

Highland Society, 92, 108–109

Highway lobby, formation of, 91–92. *See also* Centralization

Historians, 4. *See also* Macaulay, Thomas Babington

Holyhead Road, 93, 123–126

Homelessness, 21. *See also* Vagrancy

Horsepower: measurement of, 68; new roads reducing, 71

IBM's "Smarter Cities," 211

Imagination, in plotting new roads, 37

Improvement: questioned as ethos, 140; localism cutting short, 150; moral definition of, 192–197. *See also* Canals; Coasts; Rivers, improvement of; Roads; Sewers

Industrial Revolution, 16, 28

Industry, growth of, 28

Information: age of unlimited, 2; Methodists, soldiers, and journeymen carrying, 2; government designing flow of, 4; as property, 4; from state surveillance of public behavior, 19; class conflict and, 22; Parliament gathering, 49; government collection of, 80; highway lobby and, 91; about road conditions, 94; centralized oversight of roads and, 94, 99–106; Post Office collecting, 113; "small talk" filtering, 155

Information revolutions: expanding state influencing, 154; organized military travel, 156–157; Methodist circuit, 161–167, 173; artisans' tramp, 169–170; decay of, 173; physical infrastructure and, 211–212

Infrastructure: prodigious expense of, 12, 126; nation united by, 15; high cost of maintenance, 16; high taxes for, 16; as revolution in scale, 17; entitlement to, 17; as Scottish landlords' project, 17; divisive in politics, 18; Prussian, 18; under libertarian regime, 18–19; in socialist government, 18–19; necessity of, 23; as permanent structures, 26; general rules and, 51–56; expenses of, 54, 61, 139, 141; in backwaters, 61; physical magnitude of, 122–123, 125–126; tourism of, 122, 124; wealthy benefiting from, 142–143; as free-trade outgrowth, 148; modern nations made by, 198; decay of, 203–204; collapse of, 210; in China, 211; in digital age, 211–212. *See also* Broadband, as infrastructure; Canals; Coasts; Rivers, improvement of; Roads; Sewers

Infrastructure state, 4; collapse of, 199–201; in 20th century, 205

Inns: society of, 172; as social hubs of road network, 181–182

Instruments: MacNeill's "Instrument," 66; MacNeill's "Road Indicator," 68. *See also* Compass, introduction of; Maps; Theodolite

Interkingdom highway network: birth of, 80; Celtic landlords designing, 92, 100

Internet, 2

Ireland: roads lacking in, 87; London connected to, 135

Labor: hierarchical organization of, 45–46; imprisonment and poverty abolished by, 84, 88; on roads, 84, 88, 115–117; in Britain, Sweden, and France, 85; cheap and unskilled, 115–117

Laborers, as travelers, 165–166, 167–170, 172–173, 192

Labor management, roads perfecting, 16

Landes, David, 16

Landlords, 140

Libertarianism: Scotland's economy destroyed by, 18, 23; infrastructure and, in U.S., 19, 208–209. *See also* Free market; Localism

License plates, 47

INDEX

Local geography, 74
Local government: infrastructure and, 19; in 18th-century road building, 37; testimony by expert engineers and, 52–53, 61–62; roads controlled by, 85–86; corruption, 86, 96–98, 107; local grievances and, 86–87; road improvement and, 87; economic competition stifled by, 90; against road building, 96; unscrupulous profiteering and, 96–98; contradictory legislation and ill administration of, 106; expense, 107; road access protected by, 132–133; open participation in, 144; accountability and, 145; transferring costs to geographical periphery, 201; incapable of providing cheap roads, 203. *See also* Localism; Parish government
Localism, 18; emergence of, 130–133; improvement questioned by, 140; payment for roads and, 143–144; rule by experts defeated by, 146–152; nationwide triumph of, 199–200; Scotland's collapse and, 203–205
Local knowledge, engineering knowledge defeating, 61–62
London, parliamentary roads connecting to other cities, 80

Macadam, John Loudon: "General Rule" of, 51–56, 76; as expert lobbyist, 52; discipline of out-of-work poor and, 115–118
Macaulay, Thomas Babington, 12, 194–197
MacNeill, John, 66, 68
Mandeville, Bernard, 7, 88–89
Maps, 2, 9, 14, 20, 56–58, 183–187; class differentiation and, 21; parliamentary persuasion and, 27–28; plotted as opposed to surveyed, 58–59; middle-class spatial differentiation and, 183–184
Market. *See* Free market
Mathematics, 34–35, 49. *See also* Quantification; Trigonometry
Menai Straits Bridge, 5, 12, 77, 123–126
Methodists: alternative nation among, 22; as travelers, 154, 157–167, 173, 176–177. *See also* Wesley, John
Middle class, creation of, 177–179. *See also* Class; Consumption
Migrants. *See* Travelers

Migration, driven by famine or poverty, 171
Military, early roads built by, 29
Military engineers. *See* Engineers, military
Military roads, soldier traffic on, 46
Modernity, protests against, 128–129
Mokyr, Joel, 15–16

Nation. *See* Britain
National defense, infrastructure supporting, 111–112
Net neutrality, 211–212; open access to roads compared to, 132–133, 137–138, 145
Networks, social, 1; traveler safety and, 153–155, 171–178; organization of poor travelers, 172; decay of, 173

Open roads, movement for, 145, 147
Ordnance Survey, 33

Palmer, John, 15
Paper, 62–63
Parish government: open participation in, 144; accountability and, 145; defended against centralization, 149–150; gradual consolidation of, 202. *See also* Local government
Parliament. *See* Centralization; Government
Parnell, Henry, 51, 55; centralized board of roads and, 92–93, 99, 105, 108–109, 135
Participation: economic, roads promoting, 89–90; in local government, 106; opposed to "secret boards," 131, 144; local government embodying, 144–145; digital infrastructure enabling, 211
Party, political, 3, 18. *See also* Government; Participation
Pavement, 12
Pedestrians: as traffic, 13; foot pavements standardized for, 74; dangers to, 181
Periphery. *See* Geography
Permanence, of infrastructure, 46, 90
Perspective, linear, 31
Physiognomy, social contact winnowed by, 190–191
Policing: vehicle regulations, 87; traveler regulations, 191–192; public space, 207–208
Politics. *See* Centralizers; Government; Localism; Participation; Party, political

Post coaches, 15; antisociability and, 187–189

Postcolonialism: Celtic landlords designing highway network and, 93. *See also* Race

Post Office, 15; in government growth, 47; expansion of, 180

Poverty: infrastructure's promises of annihilating, 2; exile to geographical edges, 3; raking of gravel and, 4; demonization of, 22; of Scottish Highlands, 109; tolls and, 130–131. *See also* Class; Labor; Vagrancy

Prismoidal formula, 66

Private investment. *See* Free market

Private property. *See* Eminent domain

Progress: doctrine of, 2, 129, 192–197; questioned, 140

Property rights. *See* Eminent domain

Protractor, 33

Public good: roads as, 47; imagined through infrastructure, 198

Public housing, 207–208

Public space. *See* Inns; Policing; Traffic; Travelers

Quantification: in government, 48; political authority and, 56–73; of inclination, 64–65; of earthworks expenses, 68; of friction, 68; of horsepower, 68; cheapness ascertained by, 71. *See also* Mathematics; Surveying; Trigonometry

Race: Celt and English divided by, 2; roads divided by, 17; roads as leveling force for, 80, 111–112; Gaelic pride, highway lobby, and, 91–92, 100; Celtic landlords designing highway network, 93, 100; Scottish nationalism and, 108–109; English army and, 111–112

Radicals, 128, 130, 133–137, 142, 199–201; contemporary, 209

Rail, 2, 8, 202–203

Rebecca riots, 128–131

Reform, age of, 80–81, 94, 131

Region, in creation of political parties, 18. *See also* Geography

Regulation. *See* Policing; Public housing; Sewers; Speed

Rights: of householders, 78; of travelers, 83; to political participation, 88; to access roads, 134. *See also* Eminent domain; Free market; Policing

Rivers, improvement of, 13–14

Road construction: plotting of roads, 32; maps used in, 36–38; across bogs, 40–43; boulder removal, 40; leveling hills, 42–43; standardization of, 51–56, 74; analysis of foundations in, 70; standardization of milestones and guideposts in, 75; beauty of, 79–80; crooked, 96

Roads: Chinese, 2, 90; Roman, 5; Persian, 5; cross-class participation and, 8; and creation, unification nation, 9, 17, 194–197; national trade increased by, 12; in France, 14; in America, 14–15; Britain's model of, 15; Ireland, Scotland, Wales, and England as single nation and, 17; military, in Scotland, 29–30; parliamentary, 46; smoothness and speediness of, 68; cheap trade as result of, 71; expensive government linked to, 71; bringing trade to Edinburgh and Dublin, 80; centralization of, 80–88; envisioned in 1690s, 82–88; in 18th-century Sweden and France, 85; in era of local government, 87; nation integrated by, 106–120; as cure for poverty, 109–110; geography transformed by, 120; free access to, 133–138; as epitome of modernity, 192–193. *See also* Highland Roads and Bridges; Holyhead Road; Menai Straits Bridge

Rowlandson, Thomas, 179

Royal Institute of Civil Engineers, 51

Ruins, roads turning into, 23

Rule of experts, 4; economic participation and, 23; political participation and, 23; diagrams and numbers buttressing, 27; consensus, top-down governance, and, 76

Sandby, Paul, 33–34, 182

Sandby, Thomas, 33

Sanitation, 21, 81, 206–207

Scotland: war with, 29; military roads in, 29–30; roads lacking in, 87; trade participation and, 87, 108; London connected to, 120–122; corruption and, 143

Scottish landlords, 17, 91–92, 108–109

Scottish nationalism, 108–109

Secrecy, of experts, 131
Sewers, 6, 206–207
Shipping, Chinese and Indian, 5, 13
Sinclair, John, 91; Highland Society and, 91–92; *Statistical Survey of Scotland* and, 99–100; Adam Smith corresponding with, 107–108; Scottish nationalism promoted by, 108–109
Slum clearance, 6; for connection of highways, 78; as social control, 206–208; movements against, 209
"Small talk," 3, 155, 190–191
Smeaton, John, 50
Smiles, Samuel, 4, 16, 40–41, 124
Smith, Adam, 7, 88–90, 95, 199; turnpike trusts and, 107; John Sinclair corresponding with, 107–108; for tollgate abolition, 136, 148
Smith, Joshua Toulmin, 139, 144–145, 150
Snowstorms, 113–114
Sociability: etiquette, guidebooks, 183–187; maps, 183–187; 187–188; physiognomy, 190–191; "small talk," 190–191; roads increasing, 194–197; unlimited, 198
Social networks: of soldiers, 16; of military engineers turned parish surveyors, 41; of travelers, 172, 181–182
Society of Civil Engineers, 36
Soldiers: as travelers, 46, 153, 155–157, 158, 175–176; road building proposed as peacetime employment for, 114–115; links with Methodism, 158–159
Space: single monolithic, 44; military-style organization of, 45–46; regulation of, 180–181. *See also* Geography; Policing
Speed: celebration of, 124; delight of, 180
Speed limits, 180–181
Standardization: of gravel road surface, 74; of landscape around roads, 87; of vehicle axle length and horses, 87
State. *See* Government
State, military-fiscal, 21
Statistics, 46
Strangers: travelers getting information from, 20; avoidance of, 155, 187–189; interactions between, 180
Surveillance: of poor and vagrants, 19; mutual, of vehicular travelers, 87

Surveying: changes in, 30; landmarks in, 31; trigonometric, 32; instruments for, 41; parliamentary, 73; rule by experts and, 73. *See also* Mathematics; Quantification; Trigonometry
Surveyors: increase in number and availability, 36; as businessmen, 39; military-style organization of, 45–46; inability of to stand up as experts against engineers, 52–53, 61–62; corruption of, 141
Systems thinking, 7–8, 48–49

Taxes, for infrastructure, 16
Technology: unlimited progress and, 2; historians interpreting, 4; military driving, 17; role in connection, peace, and prosperity, 23; fetishism of, 25–26. *See also* Cartography, 18th-century innovations in; Compass, introduction of; Guidebooks; Labor; Maps; Protractor; Quantification; "Small talk"; Surveying; Systems thinking; Theodolite
Telford, Thomas: hero status of, 4; deathbed of, 25; transition from architect to political lobbyist, 52; as obsessive analyst, 56. *See also* Engineering; Menai Straits Bridge
Theodolite, 33, 35
Time, infrastructure related to, 46
Tollgates: protest against, 129–132; as nuisance, 132; inefficiency of, 135–136
Tourism, of infrastructure, 122, 124
Tourists, 21, 122, 124
Trade, expansion of, 18. *See also* Free market
Traffic: increase in, 12, 21; delight of, 180. *See also* Strangers; Travel
Transport. *See* Canals; Rail; Roads; Shipping, Chinese and Indian
Travel: stories and songs about, 21; expanding state causing, 154; organization through military rotation, 156–157; organization on Methodist circuit, 161–167, 173; regularization of, 179–180
Travelers: meetings between, 20; artisans as, 21; cosmopolitanism of working class, 21; Methodists as, 21; middle-class isolation and, 21; soldiers as, 21, 46; communities of, 22; on military roads, 46; tourists as, 122, 124; women as, 163, 172–174; police surveillance of, 191–192

INDEX

Trigonometry, 27, 46, 66, 71, 73. *See also* Mathematics; Quantification; Surveying

Trust: in government, 149–152; in fellow-travelers, 153–155, 171, 172–178. *See also* Social networks

Turgot, Anne-Robert-Jacques, 7–8, 89

Turnpike tolls: crooked roads produced by, 37; profiteers typifying, 98–99; best engineering practices and, 104–105; roads spreading wealth and, 107; regulation of, 119; assimilated into government bureaucracy, 119–120; smashing of, 133; private decision making, government intervention, and, 143; Utopian ideas about free roads, 147; abolition of, 201–203, 205; short-distance travelers squeezed by, 202; mounting, 204

Utopianism, 37, 88. *See also* Progress

Vagrancy, 21–22, 88, 192. *See also* Class; Poverty

Visuality: surveyors trained in, 31; trained cartographers and, 33–34; military sighting transitioning to parliamentary plotting, 58–59; in propaganda, 113–115; tourism of, 122. *See also* Built environment; Diagrams

Wade, George, 29, 33, 41, 44–45

Watercolor, 34

Webb, Sidney and Beatrice, 15, 205

Welfare, road building as, 115–119

Wesley, John, 159, 161, 164

Women, as travelers, 163, 172–174